KILLING SHEEP

The Righteous Insurgent

Mark Blackard

Killing Sheep: The Righteous Insurgent

Copyright ©2010 by Mark Blackard

First published in 2012
Edited by a team of four volunteers

All rights reserved

No part of this book may be reproduced in any form by any electronic or mechanical means, including photocopying, recording, or information storage and retrieval, without permission in writing from the author. Brief quotations may be used in literary reviews.

Library of Congress Control Number: 2012914234

ISBN-13:978-1-936956-00-5

www.markblackard.com

Printed in the U.S.A.

Warning: This book contains graphic language, graphic content, and opinions which may be considered offensive, especially to the religious conservative. Reader discretion is advised.

*For my friend, Dave White.
You were taken before your time.*

Acknowledgments

I would like to thank my friends and colleagues from the Islamic Republic of Afghanistan. It was definitely an honor to serve with them. Their actions and efforts will keep them in harm's way, long after the world loses interest in their country.

I would like to extend an honorable mention for a few fellow gringos (Americans) who are true warriors, and were not just there for the paycheck: Mullah Paul, Jackie Chan, Harry the Great, Ron, Rene, Rob C, Harry, "The Jims," and Honest Abe. I would do battle with these gentlemen any day.

Without Abe's inspiration, I could not have imagined a finished product. His encouragement and confidence drove me to start writing. Thanks for sharing your knowledge with me.

I would like to thank Sammy. He is absolutely fearless and is a true Afghan patriot. Sammy will always be my best friend and is the cocatalyst for most of these adventures. He saved my life on a daily basis. Thanks for the memories, my brother.

Thanks to Ihsanullah, with whom I shared many good times in Kabul. Welcome to America, my friend.

I started not to mention these two gentlemen, but it's not in my nature to be politically correct. I would like to say that I certainly appreciate the actions of two unnamed lazy rats. The small inconvenience I suffered as a result of their effort, betrayal, and dirty deed has only added to my credibility. It also afforded me the time to complete this manuscript. For that, I must thank them as well.

A special thanks to my good friend Smitty. You endured hours upon hours of listening to me while I read paragraphs from the book. Thank you for your time and support.

Last, but most important, I want to thank my friend who we'll call Hope. She conducted the final edit which turned a manuscript full of jumbled thoughts into a readable product. The layout, design, and typesetting are to her credit as well. It takes a special person to volunteer on such a labor intensive project. I sincerely thank you for your work.

Contents

Foreword i
Introduction 1
Journey to Afghanistan 9
Welcome to Jalalabad 31
Come to Jesus 57
Gaining Momentum 85
Islam and Beer 109
Cowboys and Indians 129
Wake Up, Mr. Bomb Maker 143
The Threat is Out There 159
Permission to Engage 171
Hitchhikers and Cellos 189
Operation Night Train 205
Typical Days 217
Stealing Heroin 235
Opposition and Suppression 245
Afghanistan Tourist Visa 257
Drunken Advice 279
Acronyms and Initials 297

Foreword

BY "MULLAH PAUL"

I first met Mark two weeks after arriving in Afghanistan. Police and military careers often draw individuals with like minds; such was the case with Mark and I. Without knowing it at the time, our paths, careers, lives, and trains of thought paralleled each other with a similarity that goes beyond words. To look at us, one would think we were complete opposites. Mark is a Caucasian gentleman from the South, and I am a first generation Mexican-American from out West. We both found ourselves on the other side of the world with individuals who varied even more than ourselves in language, customs, and religion. Yet, after several encounters with our Afghan partners sharing the mysteries of life, we discovered what made us all so similar. The underlying theme was poverty. Each had endured their own hardships and tribulations associated with this ailment.

Often enough, people in this circumstance go one of two ways: into a life of crime or into a life of fighting against it. It is no wonder why some of your better police officers are those who would have made good criminals as well. Mark's *Dirty Dozen* were the embodiment of this, as the *Biggest Gangsters* in Nangarhar Province. The Taliban would soon know what they hoped the rest of the world knew: fear. The Dirty Dozen would make sure of this.

On numerous occasions I found myself in conversation with the Dirty Dozen's upper echelon of command. On more than one occasion, I was told by Colonel G and Lieutenant Colonel A that Mark was crazy. When asked to explain, both stated, *A person who is crazy has no real comprehension of fear, or an understanding of the consequences of their decisions.* However, what they soon came to realize was that Mark was a man who had spent his whole adult life mitigating risk and looking death in the face. From his perspective, it was our job to go in harm's way to protect the average Afghan citizen. To Mark, it was simply doing what everyone was being paid to do… and nothing more.

Life looks different in the eyes of a man with no fear of death. He becomes the "threat that is out there," and we all soon became the Taliban's threat. The Taliban found themselves fighting against experienced undercover agents doing what they do best—acting like gangsters. The Taliban could no longer readily identify its enemies. They experienced the pain felt by most uniformed militaries from around the world that had fought an enemy who blended with the local populace.

I was honored when Mark asked me to write this foreword. Throughout my career as an 18-year police officer and 10-year military veteran I have read many books dealing with war and crime. But none have come close to capturing both worlds by showing the success police work can have on today's conflicts of counter-insurgency. Mark has put together a masterpiece of a raw, honest look at the war in Afghanistan from the perspective of a veteran law enforcement officer, military veteran, and civilian contractor from both the Iraq and Afghanistan theaters.

Mark often refers to the term *ground truth* in conversation. He has written the ground truth about his exploits in Afghanistan. It will entertain you, shock you, and amaze you. It is my hope that it will open your eyes to a world not yet discovered by most. Share the experience of how individuals from different facets of life, yet with great similarities, have joined together to fight against injustice. Experience the journey without having to take a risk or pay the price because after all, *The threat is out there.*

– Paul Cuellar

For evil to triumph, it is only necessary for good men to do nothing.
Sir Edmund Burke

The time has come for America to hear the truth about this tragic war. Now I've chosen to preach about the war in Vietnam today because I agree with Dante, that the hottest places in hell are reserved for those, who in a period of moral crisis, maintain their neutrality. There comes a time when silence is betrayal.
Dr. Martin Luther King, Jr.
1967

LIFE IS SHORT

LIFE IS ABOUT A SERIES OF ADVENTURES

NEVER TURN DOWN AN ADVENTURE

Introduction

The idea for this book came to me on a daily basis during my time in Afghanistan from 2009 to 2010. My frustrations and concerns as an operator, U.S. citizen, and as a taxpayer didn't seem to matter to anyone but me—especially since I was there as a contractor. I worked for a defense company who we'll merely refer to as Company X, in an attempt to avoid a threatened lawsuit (over this book). Company X is run by a bunch of retired army generals. Not a bad company to work for, but it's basically a good ol' boys club that takes care of military retirees by giving them high-paying jobs. That's not meant to shed negative light on anything, but it's the ground truth.

Company X employed me as a law enforcement advisor, of sorts. The Joint Improvised Explosive Device Defeat Organization (JIEDDO) originally funded the program I was a part of, but it eventually transitioned to the U.S. Army's Criminal Investigation Division (CID). In this program, seasoned cops from the U.S. are embedded with deploying military units in order to help combat the threat of IEDs (improvised explosive devices, i.e., bombs). I had served in law enforcement for about twelve years with the majority of my time spent as a narcotics agent. When I evaluated the position, I figured it was basically the same job I was already doing. Instead of chasing kilos of cocaine and drug traffickers, I would chase bombs and bomb makers. It was a very similar concept, so I elected to go. I was tired of working dope, the money was unbelievable, and I truly believed I could catch some bad guys and save some young lives. After two tours in Fallujah, Iraq, embedded with the U.S. Marines (another story for another time), I found myself headed to Afghanistan.

As a point of reference, *contractor* is a nasty word in this business. When military personnel refer to you as a *contractor* it's derogatory.

Contractors make more money than even the highest ranking military service members, and because of that, are viewed as criminals. Government employees are OK, though. They make roughly the same amount of money as contractors, but the military has to respect these folks because they hold "rank" according to their GS level. It was a strange experience to go from being somewhat respected as a law enforcement officer to an environment where I had no authority, my opinion really didn't matter, and the military would rather not have me around. We were each paid a quarter of a million dollars a year to be the military's advisors and subject matter experts, but we had little influence. The only thing I could do when something got screwed up was to say or think to myself, *That will definitely go in the book.*

It is hard to believe the amount of money being wasted in Afghanistan, with the absolute lack of progress and absence of productivity. It very quickly became clear to me how we could have been in Afghanistan for ten years and gotten nowhere. The best example was Bagram Air Base. Bagram was, and still is, the biggest shit hole in Afghanistan. Hard to believe since it is the main airbase where most of the supplies have to transit. Most of the troops coming and going from theater transit Bagram, as well. But after ten years, it is still the worst facility I've seen over there—an absolute embarrassment to whoever is responsible for the base. If the U.S. military were a business, it would have failed during the Afghanistan campaign. For that matter, it would have failed during the Iraq War as well, but that's another topic.

In the end, I think the U.S. involvement in Afghanistan will be almost a mirror image of the Soviet invasion, with a lot of wasted money and lives lost with little to show. The U.S. military will not fully withdraw anytime soon. Eventually, though, they will be restricted to the larger bases and merely camp out. During this manuscript I will often refer to these forward operating bases, or FOBs, as *prison camps*. That's basically what they are. Most of the soldiers who served in Afghanistan rarely left the prison camps to interact with the local people. If they did, it was for a short period of time. This was a factor in the soldiers' low morale and why many of them just didn't want to be there. One might compare a one-year tour to a one-year prison term. While a certain percentage of soldiers and Marines did engage in violent combat on a daily basis, the majority of U.S. service members never left the prison camps. The soldiers' only objective was to get back to the U.S. and break free from their confinement.

For this, I blame the U.S. military's mentality at the upper levels of supervision. U.S. military leadership has become so risk averse that they are afraid of their own shadows. This has nothing to do with the average soldier who is merely following orders. Most soldiers would love to get into the game and truly make a difference during their tours. But soldiers are severely restricted by their commanders, and commanders are concerned about getting promoted. It doesn't look good on an officer's fitness report when soldiers are killed while under his command. People have forgotten the fact that *war* is a risky business. Being in the military is risky. You know that from the time you sign up.

● ● ●

In its defense, the U.S. military is just not designed to fight unconventional wars. It is the world's best at fighting a conventional force, but the worst when it comes to dealing directly with people. They produce the best soldiers, but these soldiers often make poor police officers and diplomats when thrust into those roles. It's not the fault of the young soldier because he acts the way he was trained. Military training is centered on killing people and destroying things. Expending ordnance is the agenda. The thought process does not flex to the environment and remains rigid. This doesn't work very well during an insurgency. Unless the overall mentality changes, the U.S. military needs to stick to the mechanics of conventional warfare, get away from any business that requires caring, and definitely stay out of counter-insurgency operations.

In my opinion, U.S. Army Special Forces soldiers are the only personnel needed in Afghanistan. Their mentality and methods are as suitable as the U.S. military has for that environment. Unfortunately, Special Forces' activities are often suppressed by the *battlespace owner* (U.S. military commander) in control of the local area. Special Forces have their own command, but the strongly regimented battlespace owners who come from the regular service branches, impose their conventional warfare mindsets.

I often argued, to no avail, that the Afghan police and Afghan military are truly the battlespace owners because they are there to stay. The U.S. military commander can, at best, be called a *battlespace renter* or maybe a *battlespace leaseholder* if he was certain to remain for a year. But, since most battlespace commanders have extreme Type A personalities, no one can convince them of this concept. If you tried, they would let you know real quick, they were in charge. Although in reality, the only thing they controlled was their prison camps. Most of them just didn't get it and had no idea what was actually going on outside their walls.

...*What is an insurgent?* What the hell does that mean? The United States Department of Defense defines insurgency as an organized movement aimed at the overthrow of a constituted government through the use of subversion and armed conflict. I'm not a smart man, so I had to look up the word subversion as well. Subversion can be defined as an attempt to transform the established social order, its structures of power, authority, and hierarchy. Webster's New Compact Desk Dictionary and Style Guide defines insurgent with simplicity and stays close to the actual meaning of the word. Webster's defines insurgent as **rising up against established authority.** (However, If you need a good laugh, Mr. Katt Williams offers an interesting interpretation and perspective on insurgents.)

I wanted to reference this term from the start because it is purely subjective and based on the speaker's perspective. To me, it is merely a complicated term we use to describe a person who doesn't want to do what we, as Americans, want them to do. The words "insurgent" and "patriot" are synonymous, only differentiated by time and one's perspective.

According to the Americans, the Taliban are insurgents. All U.S. reporting refers to them as "AAF," which stands for anti-Afghan forces. But, aren't they Afghan? Maybe we should refer to them as anti-American forces. That might be the better description. Maybe we should refer to ourselves as anti-Afghan forces since we're against the Taliban.

In the Vietnam War, the U.S. military similarly referred to the Vietcong as insurgents. Who were the Vietcong? They were made up of poor, uneducated Vietnamese who did not want the Americans in their country. That sounds a lot like the Taliban in Afghanistan. The Vietcong also used similar tactics and methods as the Taliban. More importantly, they shared a similar mentality. As Mullah Paul points out in the foreword; poverty is the common denominator.

Let's discuss the Taliban for a moment. What exactly makes a person an official member of the Taliban? There's no special school that someone graduates from to become a member of the Taliban. There's no special tattoo that they get. No membership card. So, exactly how do we determine whether a person is a true member of the Taliban? Maybe we could ask him? That would be the simplest method but it's not always practical. I bring this up because the average American (including members of the military and intelligence community) thinks that any person in Afghan clothes carrying an AK-47 must be part of the Taliban. The truth is, about ninety-five percent of the Taliban are merely farmers. The hard-core extremists only make up five percent. The extremist five percent comprise the leadership and receive funding from various extremist donors, predominately from Saudi Arabia.

Saudi Arabia—now that's where the real terrorists come from. Remember 9/11? Where did most of the hijackers come from? Saudi Arabia. I often ask people at random, *How many Afghans were on the planes during 9/11?* Or, I will ask them, *How many members of the Taliban were on the planes?* The answers they give vary from two to forty. The correct answer is zero. None. No Afghans were on those planes. No members of the Taliban were on those planes. Fifteen of the hijackers were from Saudi Arabia, one was from Egypt, one was from Lebanon, and two were from the United Arab Emirates.

The funding flows to the extremist leadership, who in turn pay poor farmers to pick up arms and fight for them. The extremists pay the equivalent of one hundred U.S. dollars per month to each fighter, sometimes more, sometimes less. It is easy to recruit from poor, uneducated males who have families and no other sources of income. Most of the fighters can either choose to work for the Taliban and feed their families, or they can stay at home and starve.

Add to this the culture of the Pashtuns, who are the largest ethnic group in Afghanistan. Poor Pashtuns are very similar to poor, country people in America. Simply put, Pashtuns are like rednecks. (I can speak intelligently about rednecks because I come from a long line of them.) Pashtuns are farmers. They are hard workers. But, they are not formally educated. They can't point to Australia on a map. They can't even point to Afghanistan on a map. They are just good old country people who are firm in their beliefs. Most of them have no concept of politics, nor why the Americans are in their country. They view the American presence the same as they viewed the Russian invasion. They only know that foreigners have invaded their country and it's their duty to fight them. It's almost a patriotic duty. It doesn't get any more complicated than that, to many of these folks.

Compound the problem with the fact that Americans have been portrayed as Christian crusaders by extremist propaganda and it makes the problem worse. Now, it becomes an attack on Islam. So, if they fight for the Taliban, they get a paycheck and are viewed as patriots by many.

The young fighters are merely following in their fathers' footsteps. In the 1980's, American tax dollars paid their fathers to fight the Russians. The Russians, at the time, referred to their fathers as insurgents. Simultaneously, Americans referred to their fathers as *Mujahideen* (holy warriors), who were our allies. Again, the word "insurgent" depends on the speaker's perspective. It's just that we, as Americans, seem to think our view is always the correct one. Therefore, we label anyone who opposes American views as an insurgent, no matter what the host nation's government thinks. If the Americans call these fighters insurgents, then the Afghan government has no choice but to go along with the program and call them insurgents, too.

● ● ●

During my time in Afghanistan the world media reported their interpretation of current events, which had little to do with the ground truth. The average American could either listen to Fox News or CNN to gain a perspective on the war. While there were indeed Fox and CNN embedded reporters around the country, most news reports came from Kabul. Contrary to what you have heard or seen on the news, there was, and is, no war in Kabul. Kabul is safer than any comparably sized city in America. I will concede that occasionally a car bomb (officially called a vehicle borne improvised explosive device, or VBIED for short) will explode and kill some good people. Occasionally an attack will occur. The attackers are always quickly suppressed by the local Afghan police, without requiring assistance from the U.S. military or other coalition forces. However, the bad guys succeed in their public relations campaign because events are heavily reported in Kabul. Smoke, flame, and drama sell newspapers. News generates fear. Fear sells up-armored vehicles. Fear sells security contracts. Fear causes Congress to release funds for the military. The defense contractors get rich off of the military's funding. Fear makes money. Let me say that again, *fear makes money.*

The reporters who reside in Kabul live a relatively normal life. They stay in a respectable guesthouse that has wireless Internet, a decent restaurant, and often a nice bar. They have locals who drive them around when they need to travel. They receive information from their contacts and then throw it out there in time for the evening news in the States.

I compare reporters to the personnel who work at the U.S. Embassy. They rarely leave the safety of the embassy, rarely interact with the

Afghan people, and really have no idea what's going on in Afghanistan. They are more concerned with internal policy, sending reports back to Washington, and figuring out when their next leave period will be. I have heard that embassy personnel in Kabul get about six vacations a year.

• • •

From this introduction, you may have formulated the opinion that I'm Anti-American or against the war in Afghanistan. That is not entirely true. I am proud to be an American, but as a patriot, I'm disappointed in what I have seen and experienced at the hands of my own government. I am disappointed in the mentality of the leadership within the U.S. military. Their methods have caused undue stress on our young soldiers and placed them in harm's way, unnecessarily.

After ten years of conflict in Afghanistan we have accomplished little to nothing. How long do we intend on deploying our young people for a year at a time? The fighting in Afghanistan will go on indefinitely, with innocent Afghans being slaughtered in the crossfire. It will go on as long as we decide to stay and operate as we have. I just don't believe that we need to continue taking away years from our young service members' lives. It's not necessary. They're tired of deploying. They're tired of returning home to divorces. They're tired of not being able to drink a beer or have sex for an entire year at a time.

While Americans are losing their homes in the States, the U.S. government is shelling out money freely in Afghanistan. That money is accomplishing nothing, while breeding corruption. The corruption, however, has nothing to do with the Afghan people. If you went to any city in America and started handing out money with little-to-no oversight, people would line up for miles to take advantage of it. I blame the inefficiency and incompetence of the U.S. government for the problems with corruption in Afghanistan. They failed to maintain the proper checks and balances on the programs funded.

I compare it to the aftermath of Hurricane Katrina. The U.S. government shelled out emergency funds, and people bought new rims for their Escalades with the money. Isn't that corruption? We didn't stereotype an entire race or ethnicity as corrupt because of this, did we? So, why consider all Afghans corrupt? In truth, the U.S. government is responsible. They did not properly manage our tax money. Now, let's take responsibility for that and stop blaming others.

I don't view my opinions as derogatory, rather they are "lessons learned" from someone who isn't trying to get promoted. I'm only saying what many others cannot because of their positions and their retirements.

The content of this book is based upon my real life experiences while in Afghanistan. I am not a scholar, nor a professional author. I do not have a college degree. All of my opinions are drawn from observed reality and not academic research. Research wasn't necessary. I did time and formulated my own opinions. And, opinionated I am. However, I'm old enough at this point in my life to state my opinions without fear of the repercussions or the worry that I'm the only one in dissent.

I spent my time in Afghanistan living and working among the Afghans and not tied down to the prison camps. I learned to eat like an Afghan, act like an Afghan, and wipe my ass like an Afghan. I saw and experienced things from an Afghan perspective.

We did a lot of work while in Afghanistan and saved a lot of lives. To my comrades and I, it was all about mission accomplishment and not about the method. Some of our methods could be construed as policy violations or possibly even criminal. I don't care. I will proudly represent any and all of my actions before the American public and am certain they will support what we did. We had bombs to seize, bomb makers to arrest, and weapons caches to destroy. If we had followed the rules set forth by the U.S. military, young soldiers in our province would have died. I took the job seriously. Our main objective was to save Afghan and American lives, and that's what we did. Nothing else mattered.

We also happened to have a little bit of fun along the way. Show me a narcotics agent who doesn't drink, isn't divorced, and does not break the rules, and I'll show you an agent who can't do undercover work. This mentality spilled over into Afghanistan for a noble cause.

I hope you enjoy.

NOTE: Many names within this script have been redacted to preserve true identities for security purposes, to protect the innocent, and in most cases, to protect the guilty.

CHAPTER 1

Journey to Afghanistan

On September 11, 2009, I found myself on a U.S. government chartered flight from Atlanta, Georgia, to Kuwait. My ultimate destination was the Islamic Republic of Afghanistan. Eight years after the 9/11 attacks, I was going to look firsthand at the Taliban. Compared to my deployments to Iraq, I was much more excited about Afghanistan. As a child, I had studied about Afghanistan and the Soviet involvement there, and I still owned several old magazines from the 80's that contained articles about the Soviet invasion. To me, this deployment seemed more like a real life history lesson, rather than a war.

I had a lot of questions I wanted answered. For instance, *was the U.S. invasion of Afghanistan after 9/11 truly warranted?* I have worked for the government in some form or fashion for most of my adult life and have learned not to trust most of what they (U.S. government employees) say. Being raised as a staunch patriot, though, it took me a while to realize that U.S. government officials will say whatever they need to say to make themselves look good and keep their job security. "Ground truth" is not in their vocabulary. Instead, dog and pony shows are the accepted norm.

In particular, the one thing I couldn't understand was why—if most of the 9/11 hijackers were from Saudi Arabia—did we invade Afghanistan? Yes, Osama Bin Laden had set up training camps there, but if he had rented a farm in Idaho to do some training, would we have invaded Boise? What if he rented a cheap place in Mexico; would we have initiated a bombing campaign there? I don't think so. While I am admittedly not a scholar, I am no fool. I've read many publications and sat through numerous government training courses on terrorism, hijackings, and the 9/11 attacks. I served as a U.S. Federal Air Marshal during my law enforcement career and thoroughly studied the events of 9/11, as well as other attacks conducted by Al-Qaeda. I've read the sanitized and politically correct, *The 9/11 Commission Report*.

Another thing I couldn't understand was why we couldn't capture Osama Bin Laden—a guy over six feet tall with the most infamous face in the world. Using all of our technology, hundreds of billions of our tax dollars, and every three-letter intelligence agency, why couldn't we capture Bin Laden early on? Of course, I should have known the answer to that question. Hell, hundreds of U.S. FBI (Federal Bureau of Investigation) agents spent years trying to track down Eric Robert Rudolph, who was just one guy hiding out in the North Carolina woods. In the end, it took a rookie cop to find him rummaging through a dumpster.

What happened to the U.S. government, military, and intelligence community? I believe they have become so technology-dependent that they have forgotten how to deal with people—especially poor people from other cultures. What does a young guy with a master's degree in economics have in common with a Pashtun tribal leader who doesn't even have electricity in his home? Not much of anything. What do a bunch of college-educated FBI agents have in common with a bunch of North Carolina rednecks, who all knew where Eric Robert Rudolph was hiding? Absolutely nothing. Thus, the results of that manhunt.

The intelligence agencies have wrapped themselves up with so much internal policy, internal procedures, regulations, levels of approval, reporting requirements, and other bureaucracy, that they can't function. They can't do what they were designed to do. They are absolutely ineffective because they only look at the method and not the result. Don't agree? How about the U.S. CIA reports about weapons of mass destruction in Iraq? Somebody was a bit off on that assessment. How about the fact that it took ten years to locate Osama Bin Laden? How about 9/11? There were so many blatant clues and indicators that Al-Qaeda was planning 9/11, a high school student could have pieced together the plot.

Look at the FBI's track record in conducting investigations, as well. Between the debacles at Ruby Ridge and Waco, deductive reasoning screams incompetence. Those two situations are prime examples of Ivy-League decision-makers trying to run a simple street operation. After things went bad, they attempted to negotiate with poor, uneducated idealists, of whom they had no understanding. In reality, their best course of action would have been to go back to their federal buildings and allow the local Sheriff's Office to resolve the situation. Both ordeals amounted to a clash of cultures that ended in unnecessary violence and bloodshed.

What about the hell they put poor old Dr. Wen Ho Lee, Richard Jewell, and Dr. Steven J. Hatfill through?

Anyway, all of these thoughts were going through my head as the plane took off from Atlanta, Georgia. The military charters old DC-10s to fly personnel to Kuwait. This model has the third engine in the middle of the tail, and according to what I've been told, they are very fuel-*in*efficient. The airlines got rid of them years ago. A few cargo airlines still use them, but not many passenger carriers do. The planes were neither economical nor profitable for an airline to operate. They sold them to a private company that could pass on the expense to the U.S. government. The government doesn't care about the costs because it's not their money. It's yours. So, a fleet of old DC-10s ferry troops, civilians, and contractors back-and-forth from the States to Kuwait every day. The added fuel costs are passed on to you and me, the taxpayers. It's not rocket science. If the airlines won't use them that should be evidence enough.

Packed in the old DC-10 like sardines with weapons littering the floor, we climbed through the clouds with relative silence on board. There was no excitement on the flight.

We had just spent a grueling week at the Continental United States (CONUS) Replacement Center in Fort Benning, Georgia (or CRC as it's commonly called), where we sat through hours of useless classes and showed up for pointless formations. (The pre-deployment week basically amounted to a waste of our time and the taxpayers' money.) All deploying personnel also underwent medical examinations and received any additional required immunizations—like the Anthrax shot. We had no choice but to get the shot, but I suspect we will all develop health problems later in life because of it. I didn't trust the immunization then, and I don't trust it now—but it was a price I had to pay to clear medical. The government insists that the shot is safe, but they said the same thing about Agent Orange in Vietnam.

A trip to the Central Issue Facility (CIF) loaded us down with four huge duffle bags full of gear and a rucksack. There was no way I was taking all of that crap with me. When I was in the military twenty years ago, we went everywhere with nothing but one rucksack full of gear. It didn't matter if you were traveling for one night or one year. Now, I was loaded down with *four* bags and a rucksack. I tried to decline the stuff but was respectfully informed that I had to take it. I do have a theory on this particular policy. First, a manufacturer of a certain piece of equipment convinces the military that they have to have it. After going through the process of getting the contract, they start delivering the product. The military then forces soldiers and other personnel to draw the equipment in order to justify the purchase. It also ensures future orders for the company. This process has been repeated so many times that now, individual soldiers have to drag four duffle bags full of gear to Afghanistan. Most of the stuff they will never touch. Personnel run a real risk of hurting their backs trying to drag all of the gear around. If they lose a piece of gear, they have to shell out cash for it when they make it back to the States.

While they forced me to take the gear, it immediately went into the trunk of my car for transport to my garage in Atlanta. That's where it all stayed, in pristine condition, until I returned it a year later. Imagine the amount of fuel that could be saved on those fuel-inefficient DC-10s without all of the unnecessary weight from the unnecessary gear. Imagine all of the back injuries and hernias that could be avoided if soldiers were not forced to draw all of that crap.

The flight was quiet, and a few hours later we touched down in Shannon, Ireland, for a brief layover and to refuel the gas-guzzler. I would have liked to have had a couple of cold beers while in Ireland, but that wasn't authorized. You see, early on in the Iraq/Afghanistan conflicts, some general, in his infinite wisdom, created/imposed the so-called General Order Number 1. This order prohibits the possession or consumption of alcohol by anyone in the U.S. military or anyone serving as contractors for the military. It also prohibits personnel and contractors from having sex, possessing pornography, or preaching their faith. The U.S. Army takes this order way too far. We were on a chartered commercial flight to Kuwait but were prohibited from drinking a beer. I need to clarify, there was no ammunition on board the aircraft. Yes, there were a lot of weapons, which could maybe hurt someone if they were used as clubs. So, what good reason could the military have to prevent a plane full of civil servants from drinking one or two beers for the ride? I suppose in theory, a drunken brawl could have broken out on the plane after a soldier drank two beers, and someone could have possibly been clubbed to death with a SAW (Squad Automatic Weapon). Maybe that was the threat?

I personally believe General Order Number 1 is a direct violation of my constitutional rights. Didn't prohibition end several years ago? Plus, the plane hadn't even reached Iraq or Afghanistan. How in the hell could this order apply in Shannon, Ireland? Some of the people on that plane might not make it out of Afghanistan. They may die in that country. So, we elect to treat a plane full of adults like children? We tell them that they will be prosecuted under the Uniform Code of Military Justice if they get caught drinking a beer in the waiting area? I find this unacceptable. I think the ACLU should get involved in this particular fight. Come on—we aren't invading a country to where all hands need to be on their A-game at all times. We've been an occupying force for around a decade. Most service members don't leave the prison camps in Afghanistan. I think it would be safe to allow U.S. citizens and service members the courtesy to drink a beer in Shannon, Ireland. I'm sure they'll sleep it off by the time they reach Kuwait.

After a couple of hours walking around the gift shop area and cussing the asshole responsible for General Order Number 1, it was time to load up. We left Shannon, Ireland, in the drizzling rain. After several more hours we landed in Kuwait.

We were herded like cattle onto awaiting buses for the trip to Camp Ali Al Salem, Kuwait. Ali Al Salem houses a transient camp for personnel deploying into Afghanistan. The bad thing is, it's hot. I'm talking 120-plus degrees. The good thing is, the tents have air-conditioning. There are laundry trailers and pretty decent latrines, too. Ali Al Salem also has a McDonald's, a Pizza Hut, a Kentucky Fried Chicken, a donut shop, and a Green Beans Coffee Shop. It's actually not a bad place to get stuck for a few days, but did I mention it's hot?

As luck would have it, there was a flight leaving for Afghanistan in about four hours. It's rare to get out of Ali Al Salem that easily. I did the duffle bag drag to the waiting area and set up camp next to a reputable-looking gentleman in civilian clothes. When you are traveling alone, you have no choice but to leave your bags unattended at times. As a cop, it's against my nature to leave my belongings unsecure, but I had to hit the Pizza Hut one last time before departing for Bagram. I beat feet over to the food court and ordered a pepperoni pizza with two Cokes. I walked back to my bags and enjoyed my last slice of Americana for a while.

● ● ●

A young soldier called off names from his roster and told us to line up. We then walked to the awaiting C-17 and began taking our seats. The middle of the C-17 has four seats next to one another, like the middle of a commercial Boeing 747. It also has jump seats on both outboard sides that are twice as spacious as the middle row. These are the equivalent of flying first class, military style. Veteran travelers make sure that if they are flying on a C-17, they are at the end of the line. The flight crew loads the middle first and then directs the leftovers to the outboard jump seats. Yours truly was lumbering at the end of the line and indeed secured a coveted jump seat. Experience pays.

The C-17 took off like a rocket and we were immediately airborne. Other than having to wear body armor during the flight, I was quite comfortable and was soon fast asleep. Several hours later, we touched down at Bagram Air Base. By then it was daylight and I could see the mountains surrounding Bagram. Afghanistan was beautiful. I thought, *I'm going to love it here.*

After getting my identification card scanned, I retrieved my bags from a luggage pallet and used a phone in the terminal to call my point of contact. After about two hours of frustration, I finally reached the knucklehead. It took him another hour to show up. Within five minutes of being around the guy, I decided that he was an absolute moron. This gentleman seemed annoyed by the fact that he had to deal with me (i.e., do his job). He only had to shuttle me to the transient billeting office and drop me off, but apparently I had interrupted his doing nothing. This guy was *third*-in-charge of the program in Afghanistan at the time, and was a shining example of a contractor. (Too often in contracting work, the company only cares about putting a warm body in a position and billing the government.) After securing a temporary spot in one of the tents I was free and clear of the idiot. Not a good welcome to the program in Afghanistan, but I wasn't complaining. (This particular moron was eventually fired from his position, only to be re-hired and sent to Iraq.)

I made my way over to where the indoctrination class was taking place and hooked up with two good friends I had worked with in Iraq. It was good to see them. Of course, it's always good to see a friendly face when you're new. They had left Iraq and taken several months off before coming to Afghanistan, while I had come straight over. The indoctrination class was on a break when I walked in the room.

"How the hell are yah little buddy?" exclaimed Old Jim.

"Living the dream, my friend," I said, and gave old Jim a big hug, like brothers do.

"What the hell have you been up to?" hollered Brian, and we embraced as well.

"I've been working while you two have been sitting at the house."

We all had a good laugh and tried to catch up on what one another had been doing since we left Al Anbar Province, Iraq. A few minutes later, the orientation class resumed. I had only missed a day and was able to join Jim and Brian's class. This orientation class covered the same useless topics as every other. The instructors stood up and told the class "how things really are," but most of them had never even been off of the prison camps. *Just get on with it so we can get out of here and start working*, I kept thinking.

After class, Jim, Brian, and I deployed ourselves to the Green Beans Coffee Shop to discuss our upcoming assignment. We didn't know much about it. We would be assigned to a subordinate program under Combined Joint Task Force (CJTF) Paladin, which is the counter-IED task force in Afghanistan. The mission of the task force was to combat IEDs in theater and provide training to military personnel.

After five days of residing on Bagram Air Base I was ready to leave. I didn't care where, but I had to get away from that crowded jumble of disorganization. I've lived in poor conditions and have no problem doing so, but the Bagram latrines were so nasty—I would much rather have been at a remote OP (observation post), wiping my ass with a rock, and taking showers with a bottle of water. It would have been more sanitary. I left Bagram with a taste of disgust in my mouth for the way those young soldiers and airmen had to live. I felt embarrassed for how far we had come (or not come) in ten years of being there. Who the hell is to blame for this failure? Is there anyone in charge?

● ● ●

Old Jim, Brian, and I caught a flight to Kabul Airport where we met a couple of our counterparts, Oscar and Jim #2. After gathering all of our gear, we walked to the parking lot, looking for our transportation for the three-mile trip to Camp Eggers. When we saw our transport, our initial impression was, *Are these guys crazy?* They were riding around in Toyota Forerunners with no armor and no communications. Jim, Brian, and I went with the flow, but we were all a little concerned. You couldn't do this in Iraq. You'd die.

The drive to Camp Eggers was uneventful and took only about five minutes after we got out of the airport gates. It was apparent that Kabul, Afghanistan, was not like Fallujah, Iraq. If U.S. personnel tried to drive in a soft-skinned (no armor) Toyota Forerunner from Camp Baharia, Iraq, to Fallujah, someone would throw an RKG-3 grenade at them before they made it two miles. The bad guys wouldn't stay and fight, but they would damn sure throw a grenade.

No, this definitely was not Iraq.

As we pulled into the parking lot, Jim informed us we'd have to stay in an old tent for a few days until they could secure better accommodations for us. No problem at all. We wrestled our gear into the tent and Brian immediately grabbed an open bottom bunk. That's prime real estate in a tent. After he staked out his claim, only one set of bunk beds remained. Since Old Jim is getting on up in years, I told him I would sleep on the top bunk. It was crowded in the tent, with about twenty gents living there at the time, but our immediate complaint was the mosquitoes. A ton of them were buzzing around inside. The damp location obviously suited the little bastards.

After securing our bags, Jim gave us a tour of Camp Eggers. He informed us that a ton of lieutenant colonels resided on the base, along with ninety-three colonels, and even a couple of generals. Therefore, personnel needed to be on their best behavior and always look presentable. Anyway, we walked into the camp and I was in shock. Camp Eggers looked like a college campus.

The U.S. military had taken over several city blocks to carve out the camp. The area was well lit, and the streets were clean and paved—there was no dust. There were women walking around everywhere. A quick tour revealed that the campus had a pizza place, a Thai Restaurant, a Post Exchange (PX), a nice gym, a Green Beans Coffee Shop, wireless Internet, a rose garden, and a spa. *You've got to be kidding me, a spa? I thought we were in a combat zone.* I couldn't believe there was a spa on the campus with beautiful girls from Kyrgyzstan who, for twenty dollars an hour, would give oil massages. This was not exactly what I had in mind when I envisioned Afghanistan, and it was definitely a far cry from Fallujah. The place was surreal.

Jim escorted us to the smaller of the two chow halls on the campus; it was called "The Goat." As we walked up to it, Brian made sure he was the first one through the door.

"Damn, you've always got to be the first one in line," said Old Jim as we walked in.

"That's right. You snooze you lose."

Brian is a great guy and a good friend. But, he's the type of person who always has to be first in line, no matter what. He always has the latest gear and the best equipment, too. If you have something he doesn't, you can be sure he'll be online ordering something better. It's just his way, but we still love him like a brother. If you walk into a place and there's only one chair to sit in, you can be sure Brian's ass will be in it. If you caught a fish this long, then he's caught a fish even longer…

The accommodations on the campus continued to amaze me, and I loved The Goat. The eating area is actually in a tent, but it's nice. The LCD televisions hanging on the wall were blaring the latest news, and the food was good, too. The cozy atmosphere felt like a quaint little café, rather than an institutionalized dining facility.

"If you like that, just wait until lunch on Fridays at the other chow hall. They serve the biggest burgers you've ever seen, fresh off the grill," Jim advised as we walked out.

● ● ●

The day after our arrival, we met with Jim and Oscar to find out about our mission. We were part of a program within CJTF Paladin called the Investigative and Surveillance Unit, or ISU for short. ISU partnered with CJTF Paladin, the Afghan Ministry of the Interior (MOI), and the Combined Security Transition Command – Afghanistan (CSTC-A). In essence, the program took teams of fifteen Afghan police officers and paired them with two of us from the U.S., in order to conduct counter-IED operations. We would work undercover with the Afghans and advise them on operations and investigations. At the time, there were plans to stand up ten such teams throughout the country.

Now, this sounded interesting. The concept was definitely up my alley. The teams would work out of rented off-site locations (OSLs) in the local community.

…Let's discuss terminology. There were certain terms we couldn't use because the folks in the intelligence community would cry, and we would get in trouble with the military.

We couldn't refer to the locations as "safe houses." There, I said the words—safe houses. Holy shit! If I had used those words during my tour I would have been fired and the program would have been shut down; people would have been indicted. That's not an exaggeration. We absolutely could not call our safe houses, safe houses. We had to use the term "off-site location." So, from here on, when you see the acronym OSL, it should translate in your mind to a safe house. This is one of the many examples of why the mentality of the U.S. intelligence community leads to failure. That community is more *concerned with semantics and policy than mission accomplishment. Never mind dealing with the bad guys, but if some nasty contractor uses the term "safe house," then they'll investigate. The bottom line is, who cares what you call it? We were operating out of rental houses.*

CSTC-A provided the funding for the safe houses and for the vehicles. A team typically had three Toyota Forerunners and two Toyota Corollas. We bought them used, so they would blend in. This was in contrast to the Afghan National Directorate of Security (NDS). They are the equivalent of our Central Intelligence Agency (CIA) and were reportedly mentored by CIA personnel.

...Oh shit! I used the letters CIA. That is something else you cannot say in this business. While our Central Intelligence Agency is officially called that, you can never use their name. You can't mention them in any reports nor in any conversation. They will cry about it because they feel as if they're too secret to acknowledge their own existence. You have to refer to them as OGA, which stands for Other Government Agency. Even then, you have to be wary of saying those letters. For this book, I'll call them the CIA, because that's who they are.

Anyway, the NDS officers drove brand new Toyota Hilux pickup trucks. Everybody knew who they were. Sure, it was nice to have a brand new vehicle that wasn't prone to breaking down, but it didn't help them accomplish their mission. We would end up dealing with vehicle issues all the time, but at least the bad guys had no idea who we were.

The members of the ISU teams were handpicked from various law enforcement backgrounds, such as the counter-narcotics or counter-terrorism fields. They were to be experienced. They would have all the necessary communication equipment, rifles, pistols, and vehicles.

The way I looked at it, these teams would operate much like a drug task force in the United States. They would run informants, conduct surveillance, conduct undercover operations, execute search warrants, and make arrests. It was exactly what I had been doing for years. The basic elements of police work remain the same, no matter where you are. The geography, specific laws, and criminal procedure can vary drastically, but putting bad guys in jail is the common goal. I was ready to get going.

After a few days of living at the Mosquito Inn, Jim informed us he and Oscar had secured two rooms. They had worked a drug deal with some Special Ops folks, who agreed to let us use their quarters. The agreement was subject to termination at a moment's notice.

...In combat zone jargon, *a drug deal refers to an agreement made between two parties that is often outside of policy and the normal procurement channels. The best accommodations are always obtained via drug deals.*

This meant that two of us would have to share a small room, about the size of a broom closet, while one lucky guy got his own place. No problem. At least we'd be out of the tent.

After breakfast, Jim showed us the first room, which was next to the chow hall, in the basement of one of the houses that served as an office building. It had a set of bunks, a desk, and a small television that was mounted on the wall. It would definitely be cramped with two people in there, but it had a wired Internet connection. It would be outstanding to have some communication after hours.

We left to check out the second room, which turned out to be a suite. It had an air-conditioner, a nice desk, a wall locker, and a television. Next door was an office with a refrigerator, computers, and printers. Plus, whoever got this room didn't have to bunk with anyone. It was a very convenient setup. Like children, we were all trying to figure out how we could jockey the suite for ourselves.

"Let's draw straws at lunch to see who gets the suite," I said.

As this was the only democratic process any of us could think of, we agreed. When lunchtime came around, we headed to The Goat for some chow and to roll the dice.

I took three straws and cut one off so that it was shorter than the other two. While we were being democratic about the situation, I was still praying I came up with the short straw.

I held the straws and let Old Jim and Brian each pick. Jim went first and came up with a long straw. Good news for me. Now I had a fifty percent chance that the suite was mine. Brian grabbed a straw and came up with the short one.

"Damn it, Brian!" exclaimed old Jim. "That's just like you to win the best room."

Brian was grinning from ear to ear and then broke out into laughter. He loved being the winner, and he knew Old Jim was absolutely correct in his description.

"Brian's that guy at the dinner table who always has to have the big piece of chicken," I said. "If you bring a bucket of chicken home, he'll dig through the whole damn thing to make sure he gets the biggest piece. Matter of fact, from now on we're going to call you the big piece of chicken, or The Big Chicken for short."

Old Jim agreed with me saying, "That's an appropriate name for his ass, The Big Chicken it is."

Brian is about six-foot-two, so you would assume he was given the name because of his stature. Now you know the real reason.

Jim and I ended up bunking together in the broom closet while The Big Chicken relaxed in the Hilton suite. I quickly signed up for Internet service and was soon enjoying some connection to the outside world. Since Old Jim snores like a chain saw, I'd stay up until the wee hours of the morning messing around online. That way, when I did climb in the rack, I was certain to get some sleep. It worked out fine.

I think The Big Chicken felt kind of guilty after his first night in the suite, knowing Old Jim and I were cooped up in the broom closet. He invited us to come over and watch a couple of episodes of *The Family Guy*, after hours. I had never watched that show and had no interest in viewing, what I originally thought was, a childish cartoon.

"You guys should come on up around eight tonight. I found a treasure in the refrigerator," claimed The Big Chicken.

He went on to explain that the Special Ops folks had left quite a stash of Corona in the fridge. *Were they covered under General Order Number 1?* Now he had my attention. He also had some microwave popcorn we could cook up.

"What time did you want to get together?" was my only response. I don't think I've ever turned down an offer to drink beer in my entire life.

Every day, life on the campus seemed to get better. I wasn't excited about watching cartoons with The Big Chicken, but I would certainly show up and help him drink free beer. For the next few nights, we all congregated to watch two episodes of *The Family Guy*, drink somebody else's Corona, and eat their microwave popcorn. It was good living. I also laughed my ass off, especially at the episode with Bill Clinton. Now, I watch that show any chance I get. If it weren't for The Big Chicken feeling guilty about me and Old Jim living like sardines, he probably would have enjoyed all those Coronas by himself. I'm almost sure of it.

...Now, I have to tell you. The suite sat across from a large room that served as the chapel. I'm not sure any real worship took place there, but people seemed to congregate on a nightly basis. While the three of us watched inappropriate television shows, drank beer, cussed, and cut up, the "church" held services across the hall. However, there wasn't much worship going on. They had a band set up and it sounded like a rock concert every night. No one was singing "Jesus Loves Me," either. Instead, I kept hearing

songs from American rock bands, like Pearl Jam. I don't seem to recall those types of hymns from my childhood, while sitting on the third pew during the Sunday morning service. Therefore, I didn't feel a bit guilty about our activities. I won't even mention the number of times I caught some late night "worshippers" sneaking into the room for some more violations of General Order Number 1. It didn't bother me a bit because sex should not be illegal between two consenting adults, even in a war zone.

In the borrowed office where we congregated to watch movies, the Special Ops guys had posted a picture of themselves with the caption "The Three Amigos." The picture showed snowy mountains in the background—it had obviously been taken during the winter. All three of the gentlemen wore what appeared to be, traditional Afghan blankets made into ponchos. The Big Chicken thought they were the coolest thing and insisted we add them to our wardrobes. I'll have to admit, Old Jim and I thought they looked pretty cool, too. We would look like the Outlaw Josey Wales—Afghanistan style. The Big Chicken took the initiative. He took the picture over to the tailor on the campus and reported that the tailor would make them for fifty U.S. dollars. The blankets were thirty, with an additional twenty for the tailoring. Not a bad deal. We had three of them ordered. We would definitely look like locals with those new ponchos—or so we thought.

● ● ●

Our two weeks with Jim and Oscar were a crash course on how to run a team, and we had a lot to learn. For instance, one morning we loaded up to go to the Kabul team's safe house. We met in the campus parking lot and donned our gear. At the time, Old Jim, The Big Chicken, and I elected to wear our body armor and Kevlar helmets. (We still had Iraq mindsets.) Then, as we were about to depart, I volunteered to ride in the hatch and serve as a rear gunner. Looking back on that now, I must have looked like an idiot.

Nothing happened on the twenty-minute ride to the safe house, except for encountering some terrible traffic. When we pulled up, the house resembled a small palace. With a honk of the horn, a couple of Afghans wrestled open the heavy metal gate and we pulled inside.

Oscar stepped out of the Forerunner. With his dark hair and complexion, he could have passed for an Afghan, except that he stood about six-foot-three and was built like a linebacker. You can't hide stature.

As a handful of Afghans entered the courtyard from the house, Oscar met them each with a hug, repeating *As-Salamu Alaykum, As-Salamu Alaykum*, as he went down the line. The Afghan agents greeted Oscar like a celebrity.

> *...Oh shit!* I used the word "agent." While these gentlemen were the equivalent of federal agents, we were forbidden to call them "agents" in our reports. The U.S. military ordered us to refer to them as "officers." Imagine if someone in the States referred to a special agent with the FBI as an "officer," rather than a "special agent." "Officer" wouldn't be his proper title and it could be construed as a subtle lack of respect. The Afghans were federal agents, and therefore, should have been referred to as such and afforded the proper courtesy by the U.S. military.

After fifteen minutes of greeting, we settled onto the rooftop patio. While it had a beautiful view, I felt apprehensive about sitting on the roof. The safe house sat among a sea of other buildings and snipers could be hiding in a hundred places. I was ready to go to heaven, but I didn't want to give my life away for free.

A survey of the area revealed that almost every house on this particular street looked like a television station. Each house sprouted more than a dozen antennas, satellite dishes, and towers. It was obvious who owned them. Oscar began to point out the various intelligence agencies that lived there.

"There's some Americans, there's the Iranians, there's the Pakistanis, there's the Brits." Oscar pointed them out one by one. "The property dealers know them all. They talk amongst one another, so all the Afghans know who lives where. That means we know as well."

Oscar went on to explain that since our safe house stood nearby the other intelligence agencies, we could benefit from their security. Our teams weren't budgeted any security personnel, so the agents had to take turns pulling duty at night. That only left two or three people to defend the place in case of attack. But since every intelligence agency on the street had privately contracted security, sitting outside twenty-four hours a day with AK-47s, our safe house was safe, too. All for free, and especially if you bought their guards lunch every now and then. Hell, they'd watch our place closer than where they were posted.

We spent the entire day at the safe house and I watched, listened, and learned how the team operated. Their pace was slower than I would have expected, but they seemed to talk on their phones a lot, agents came and went, and they seemed to be doing business. I began to get a feel for what was going on.

While the house had four floors and about twenty rooms, all of the Afghan agents (minus the leadership) congregated in a single room that was lined with pillows and cushioned mats. It had no furniture, as was typical in Afghan homes. A teapot sat in the middle of the floor with freshly brewed tea. The agents gathered around the teapot to drink and talk. Personally, I thought that instead of sitting around, they should have been out in the streets working. They should have been conducting surveillance on some bad guys or jacking up informants for information. But since it wasn't my show, I just sat back and took notes.

This became the routine during our time with Oscar and Jim. While we glimpsed the concept, we never actually went on a mission with those guys. I'm well aware that in the drug business the good guys can't just go out and buy a case, but at the time I had hoped to see one of their operations firsthand.

As we continued to pick Jim's brain about administrative and logistical issues, I had one concern. How much money were we going to be budgeted to pay the Afghans' informants? This was not my first rodeo and I knew that funds to pay informants were always overlooked. The bosses expect you to work cases without money. They never understand that you need money to operate because most of them have never been drug agents.

Whenever I asked about the money, Jim would respond by saying that CSTC-A had pumped funds into the Afghan Ministry of the Interior (MOI) to pay informants. There was a process in place, and we could request funds through Afghan channels. As he explained the long, inefficient process, I detected some deception in his voice. He attempted to avoid the subject every time I brought it up. Nevertheless, I continued to ride him about it, and in the end, discovered that we advisors had no money to pay informants. That was going to be a problem.

In the United States, most individuals who become confidential informants (CIs) have been arrested and are facing criminal charges. They either bond out of jail or we get them out, so they can work for law enforcement. In exchange for their cooperation, they receive "judicial consideration." In simplistic terms, they're working to keep from going to jail because we have them bent over a barrel. They have no choice other than to cooperate with law enforcement, or else they go to prison. In Afghanistan, the criminal justice system doesn't work like that. I suspected, and soon confirmed, that the only reason people in Afghanistan cooperate with law enforcement is for money. They're predominately poor and are largely motivated by financial gain. When they cooperate they take a real risk; both for themselves and their families. A risk that could end in death. They should be compensated for their efforts.

I couldn't understand why the U.S. stood up these teams without funds to pay informants. We were all experienced law enforcement agents who dealt with informants every day in America, from interviewing them, directing them, and *paying* them. Whose bright idea was it to leave that off the budget? It was like paying some instructors a ton of money to teach truck driving, without supplying any fuel.

The other issue, Jim promptly explained, was that we had to maintain the façade that we, as advisors, did not "run sources."

...*Oh shit!* *I said the word "source." Somewhere, as you read this, there are intelligence folks having a meeting. Did he really say that word?*

Jim explained that only properly school-trained military human intelligence (HUMINT) personnel could run sources. According to the company line, we didn't run sources; instead, we advised Afghan police officers who worked with police informants. And we weren't allowed to deal directly with the Afghans' informants. If we did, we had to write the report saying that the information came from our "law enforcement contact" (LEC) instead. Plus, we didn't write "intelligence reports." We wrote "criminal information" reports.

Now, this all seemed a bit ridiculous. The taxpayers of the United States were paying me a quarter of a million dollars a year to advise the Afghans on how to catch bad guys. They were paying for, and understandably should expect, results. I had been doing the job every day for over a decade in the United States. I'd done the job while operating under the scrutiny of supervisors, the media, defense attorneys, prosecutors, and the courts. That's why I was hired. I am a subject matter expert on running informants and conducting interviews.

However, since I was in Afghanistan as a contractor, the United States military immediately classified me as an idiot. So, why would they hire me and pay me all that money? I worked for the U.S. military, so why wouldn't they expect me to do what I'm an expert at doing? The whole concept defied logic and just didn't make any sense to me. If I pay a landscaper, I expect him to cut my grass and not just hang out in the yard. But, who wouldn't?

...Here is an opinionated side note to contrast law enforcement with the military. *The military trains ninety percent of the time, for what they will do in reality about ten percent of the time. Law enforcement is the exact opposite. Law enforcement trains ten percent of the time, for what they actually do ninety percent of the time. The military is ninety percent training and speculation, while law enforcement is ninety percent experience and reality. A year of experience is worth ten years of training.*

Let me throw out some food for the lawyers to fight over, since I'm sure they'll weigh in on the subject. What exactly classifies a person as a "source" anyway? If a man walks in off the street to Campus Eggers and wants to tell me where a bad guy is hiding, does that make him a source? Should I tell him to be quiet and hold his information until I can find a school-trained, properly authorized, twenty-year-old soldier who can take down the information? What if I am drinking tea with a local Afghan who runs the barbershop, and he tells me where a bad guy is hiding? Is my barber now a source? Thus, making it illegal for me to converse with him any further since I'm a contractor? What if he's a really good barber? Can I still go get a haircut as long as I don't talk to him? Let's establish some case law on this subject because it's hard as hell to find a good barber these days.

This topic builds on my opinion that the U.S. intelligence community, as a whole, is ineffective because of the mentality ingrained in their sub-culture. I don't blame the individual agents (military or civilian) trying to do their jobs and make a difference. They were, and are, just as affected by the internal red tape as I was.

The main focus within the U.S. intelligence community is on whether or not policies and procedures are being followed, rather than whether mission objectives are being accomplished. They only care about the method and not the results.

I'm not saying that it's OK to waterboard someone, resort to physical violence, or use painful interrogation techniques to accomplish a mission. I'm a cop and have never been party to that type of activity. I would have gone to jail if I had ever abused a suspect. *Although, you can apparently get away with it at Guantanamo.* I just mean that there's so much control involved with merely talking to someone, that the system is broken. That's why it takes so long for information to get out—it has to be reviewed and re-reviewed to make sure it doesn't violate any internal bureaucracy. God forbid the possibility that good information could come from a police advisor who talked directly to a source.

You know, speaking of Guantanamo, it's kind of funny that I could have gone to jail for running sources (i.e., merely talking to people) in Afghanistan in order to save American lives. But had I worked for the CIA, I could have *legally* waterboarded people in Guantanamo with no problems. I guess if you actually work for the U.S. government, it's OK to break the law and torture people. However, if you're just a nasty old contractor, it's NOT OK to *fuck* with bureaucracy. Even the average street cop can get a confession out of the most hardened criminal while he's in handcuffs in the back seat of a police car. Why? Because that's what he does for eight to ten hours a day, almost every day of his life. It is because of experience. No need to break out the waterboard from the trunk of the police car. No need for sleep deprivation. No need to study your approach. I'm not real sure what an approach is, actually. That's just a term I've heard used by the military interrogators to describe how they are going to question someone. Cops just talk to people. Since that's the majority of what they do, they're good at it. They don't have the luxury of time to be able to plan an "approach" to an interrogation.

● ● ●

In the end, I realized that I'd have to get creative and find a way to pay their informants, while keeping myself out of trouble. After about a dozen discussions on the topic, I decided to drop the subject with Jim. Besides, he had other problems to tell us about.

He explained that another major issue involved auto repairs. The undercover vehicles were purchased used, in order to blend in, and they needed routine maintenance—*but* their repairs weren't covered under any existing contracts. How and why wasn't this addressed? Jim told us that someone was working on it and that we should do the best we could with what we had.

Fuel for the vehicles was another issue. All the teams had problems obtaining fuel. The Afghan supply chain was screwed up and there was no way to order it. *Another logistical issue that would inevitably require a drug deal to overcome.*

And another thing. The agents carried Russian-made Makarov pistols, which fired 9x18mm rounds. These are just a bit smaller than the common 9x19mm rounds used by the U.S. The problem was that there was no way to procure ammunition for these weapons. No one carried the rounds. You couldn't order them. You couldn't buy them. Whatever the agents had was all they were going to get. The Soviets had reportedly designed this round to be smaller on purpose, so that in case of war, NATO forces couldn't use seized Soviet ammunition against their own forces. It worked in this instance, in reverse. We couldn't use U.S. ammunition in *their* pistols.

How in the hell do you take these guys to the firing range to get some training if they only have a finite amount of ammunition? What if they got into a firefight? How would we replenish their duty ammunition? Who came up with the bright idea to equip Afghan undercover agents with weapons you couldn't obtain ammo for? It had to be the U.S. military—because it made no sense. The Germans definitely didn't have anything to do with it. It had to be our own people who screwed this up.

The agents, if they were lucky, received Czech Republic–made VZ-58 rifles, which are basically knock-offs of the coveted Russian-made AK-47 (or AKM) assault rifle. I would later learn that the Afghans had absolutely no faith in these weapons and would leave them at the safe house instead of taking them on missions. They didn't trust a VZ-58 to fire a single round, and if it did, they were certain it would jam on the second one. So, why carry the damn things? They would rather depend on their trusty Makarovs and their limited number of rounds. At least the Makarovs actually fired.

The basic synopsis was that we had no money to pay informants, no money to repair vehicles, no ammo for the pistols, no fuel for the vehicles, and we were equipped with generic rifles. It wasn't exactly a recipe for success.

On top of all that, the only incentive for an agent to join our unit was a single pay-grade promotion. For example, if the agent was a captain, he would receive a promotion to major. That was the only financial incentive of joining a unit that conducted undercover, high-risk operations. The agents' average salary was only about two hundred U.S. dollars per month, by the way.

Fortunately, according to Jim, we had the power to hire and fire. While technically the Afghans had to make this happen, we had the unofficial authority. This was important. I had seen some agents who I considered dead weight while visiting the Kabul safe house. I've been in the business long enough to know who's motivated and who's not. I began to refer to any agents in the program who were non-productive, as *Tea Drinkers*. Basically, all they did was show up for work and drink tea —and nothing else. I would definitely identify and weed out any *Tea Drinkers* on my team. I have no tolerance for freeloaders.

Our time in Kabul went by quickly, and it soon became time to head to our own areas of operation. Jim had identified three different locations for us: Gardez, Mehtar Lam, and Jalalabad. I like to be the captain of my own destiny when possible. In the evenings, I did a little bit of research and decided to vie for the City of Jalalabad in Nangarhar Province. First, Jalalabad was the warmest of the three locations. (I hate cold weather. I would rather be in one hundred degree heat than feel the least bit chilly.) Second, Jalalabad sat on the main route between Torkham Gate (i.e., the border crossing with Pakistan) and Kabul, which made it a major smuggling route for goods coming in from Pakistan. There would be plenty of bad guys bouncing back and forth and there should be enough action and work there to keep me busy. So, I started jockeying to go to Jalalabad, or J-Bad for short.

Another factor was that the guy I'd be replacing in J-Bad had total freedom of movement. He had his own team vehicle and could come and go as he pleased. I didn't want to end up at a prison camp with a battlespace renter who didn't allow contractors to leave the wire.

After several days of insisting to everyone that I was going to J-Bad, Jim finally wrote my name on his whiteboard, confirming my destination. While it was indeed a strong-arm tactic on my part, I certainly appreciated Jim bending to my asserted "request."

Finally, it was time to leave the campus and get into the fight. We made one more trip to the Green Spa and got a nice massage from the Kyrgyzstan girls. We also devoured one last slice at the pizza place. The next day, Old Jim and I found ourselves at a landing zone waiting for a Huey to take us east toward Pakistan.

CHAPTER 2

Welcome to Jalalabad

When Old Jim and I began planning our trip to Mehtar Lam and Jalalabad we assumed we'd have to travel via military air. However, The Big Chicken had done some research and found out about Molson Air. Their official name was Canadian Helicopters, but everyone just called them Molson, after the Canadian beer. Their fleet consisted of a bunch of old UH-1 (Huey) helicopters, painted blue and white. They were piloted by Canadian civilians, who the U.S. military had contracted to fly personnel around Afghanistan. The flight was free for the passengers.

Let's analyze this for a minute. We were supposedly in a combat zone and the U.S. military had plenty of rotary wing aircraft, but for some reason, they couldn't transport their own people around the country. Instead, they contracted a private company that brought unarmed Vietnam-era aircraft with unarmed civilian pilots, to fly combatant troops around Afghanistan. *Wow.*

As we waited at the landing zone (LZ) on Camp Phoenix, located on Jalalabad Road a few miles from Campus Eggers, I began to wonder if they would take us with all the bags we had. Right on time and like clockwork, we heard the *whop whop whop* of the rotors in the distance. A few minutes later, the Huey sat down on the LZ and a young man stepped down from the co-pilot's seat. He wore a pair of old blue jeans and a ragged white t-shirt.

"You guys ready to go?" was his only question.

"Hell yeah, you got room for all these bags?" I asked.

"No problem," he responded, as he grabbed one of the heavier bags and wrestled it into the tail section of the helicopter.

Old Jim and I climbed aboard and strapped in. Two young soldiers were already in the bird along with the pilot, who was an older gentleman.

He was also wearing an old t-shirt and jeans. The two pilots reminded me of a couple of hippies driving an old VW van. They were right out of the movie *Air America* with Mel Gibson, which was based on the CIA-run airline in Laos during the Vietnam War.

The Huey took off toward the East and pretty soon we had the best view anyone could ask for. The mountains below were breathtaking. *How could anyone use this beautiful country as a battlefield?*

We crossed into Nangarhar Province and were soon circling Jalalabad Airfield, which is also known as FOB Fenty. The bird touched down without so much as a bump, and then the co-pilot helped us drag our bags out of the aircraft. We thanked him and the Huey took off, leaving an abrupt silence in its wake.

That was the most efficient, hassle free travel I have experienced since being in this line of work. There was no disorganization involved—no briefings, no identification card scans, and no baggage regulations. Just get on the bird and go.

*...I want to say this about **Molson Air**. Those guys are more dependable than the U.S. mail. In Afghanistan, they flew during rain, sleet, snow, or shine, seven days a week. They only stopped if the pilots logged too many hours. You could set your watch by those guys. A couple of months after I first flew with them, I heard that small arms fire (SAF) hit a Molson Air helicopter, and it had to make an emergency landing near one of the prison camps. A young soldier was reportedly shot through the leg and had to be transported to Bagram for treatment. Molson Air didn't even pause their operations. They drove on with their mission. While Molson was definitely an asset to have, I still question why the U.S. military couldn't accomplish the same objective? Go to any prison camp that has an airfield and you will see a plethora of military aircraft sitting idle next to the runway. Why couldn't the military transport personnel with their rotary wing aircraft? At least their aircraft were armed, while Molson's were not.*

As soon as the helo departed, we were greeted by a fellow advisor named Rene, and a young female soldier.

"Welcome to J-Bad," Rene said, as he shook my hand.

"Glad to be here, my friend."

Rene is a Hispanic gentleman from California, about forty-five years of age. He's tall, has dark features, and at the time he was sporting a beard. He was able to blend into the Afghan community with ease. Rene worked at one of the prison camps about a half-mile away from the airfield, known as Camp Hughie. He was filling in for one of our counterparts at FOB Fenty who had gone on vacation. Rene and I would quickly become good friends.

"I've already got you squared away with a room on the Paladin compound," Rene explained, as we loaded our gear into the Forerunner.

We loaded up and traveled the perimeter road that circles the airfield and snakes by the numerous compounds within the FOB. Some compounds had their own security forces. These obviously belonged to the CIA and Special Ops groups that had ample funding. (Maybe they were concerned about an imminent attack from inside the FOB as well as from the outside, thus the need for the extra security.)

Finally, we pulled up to the Task Force Paladin compound, which was a nice place as far as battlefield conditions go. Four hard-structure buildings faced one another, enclosed by a fence. There was a small courtyard in the middle. The layout formed sort of a mini-compound and it offered some privacy and security. A rickety set of wooden stairs clung to the side of one of the buildings. According to Rene, the crew gathered on the roof to smoke cigars on a nightly basis. It had a great view of the mountains, as well.

Rene ducked into one of the buildings, which I later learned contained the tactical operations center (TOC). He soon emerged with the first sergeant, who showed us to our rooms.

The rooms turned out to be prime real estate in one of the hard-structure buildings. The walls didn't reach the ceiling which actually made them more like cubicles. Each room had a desk, a dresser, and a wall locker, in addition to a makeshift bunk. This was good living on a FOB. It's a totally different dynamic when you have to share quarters with other grown men. There's no privacy whatsoever, and the other guys inevitably leave their belongings all over the place. Plus, you can never sleep because someone is always coming or going. I was definitely thankful to Rene and the first sergeant for accommodating us.

After securing our gear, Rene gave us a tour of FOB Fenty. The gym stood immediately next to our building. Personally, I could care less about a gym, but I'm one of the few. Most servicemen and contractors spend more time in the gym than they do working. If you don't believe me, just go to the PX and see how many shelves are dedicated to steroid powder, shakes, and power bars. I came to hate this particular gym, especially. Almost every night some idiot would take a set of barbells to the sidewalk outside the gym, which was about three feet from our rooms. Said idiot would pick up a barbell, pull it up to his chest while letting out a loud yell, and then drop it onto the concrete from chest height. The barbell would hit the concrete with a thud, and then he would repeat the cycle, creating a ruckus that sounded like, *"arrrghhh!"*... *BOOM! "arrrghhh!"*... *BOOM! "arrrghhh!"*... *BOOM!* The "arrrghhh" sounded like someone shitting a strand of barbed wire, and the BOOM sounded just like a 107mm rocket hitting the ground at a distance. We never knew if we were being hit with indirect fire (IDF) or if some genius was just working out on the sidewalk. Now, I didn't want anyone to get hurt, but I often wished that a rocket would hit the gym—when it was vacant, of course—and burn it to the ground. If I end up suffering from posttraumatic stress disorder (PTSD), it will probably be from the constant bombardment of barbells on concrete, caused by a few gringos with nothing better to do.

As we walked toward the chow hall, we passed the Green Beans Coffee Shop. It was the only real amenity on the camp. The Green Beans operated out of a single shipping container and there was really nowhere to congregate. Just get your coffee and go. We then toured the laundry and the Post Exchange (PX).

Rene explained that the prison camp used to have a Pizza Hut, but the camp commander had done away with it. (This was long before General Stanley McChrystal decided to get rid of all the Pizza Huts in Afghanistan.) Apparently, the restaurant stood just outside the Brigade Headquarters. One day, reportedly, the Brigade Commander (not sure who, but obviously a full-bird colonel) walked out and saw nothing but fat, overweight soldiers standing in line for pizza. He was so disgusted, the story goes, that he immediately evicted the Pizza Hut from the camp and tried to evict the Green Beans Coffee Shop as well. While the Pizza Hut was gone for good, the Green Beans didn't go anywhere. (I'm not sure what it is about the Green Beans, but apparently, they have friends in high places; or low places. Depends on how you look at it, I guess.)

What reason could that particular colonel have cited for removing the Pizza Hut from FOB Fenty? If he had concerns about health and readiness, then it was a flawed decision. The official chow hall has a fast food bar. For two meals a day, seven days a week, soldiers can eat pizza, chicken wings, cheese sticks, chili, cheese sauce, hamburgers, and hot dogs—all for free. Did he really accomplish anything by removing the Pizza Hut? Sure, he lowered the young soldiers' morale. The troops who lived in the sticks looked forward to that occasional visit to Pizza Hut. Maybe it made that commander look like a firm leader on his fitness report because he made an unpopular decision. (That will get him promoted to general for sure.)

After our brief tour, we enjoyed a good meal at the chow hall and then called it a night since we had been traveling around all day. Plus, Old Jim had to get on to Mehtar Lam the next morning, while Rene and I had to start handling business in J-Bad.

I slept like a rock that first night on my makeshift bunk. Those old mattresses were just right and there were no assholes congregating on the sidewalk dropping barbells.

The next day, after we got Old Jim on a Molson Air flight up to Mehtar Lam, Rene started spinning me up on the situation with the Jalalabad team. He explained that the previous advisor had to suddenly return to the United States after serving for over a year. He had built a lot of logistics with the team, but his efforts had yet to show much operational progress.

The team had one interpreter; a local guy from Jalalabad. We were also allotted an interpreter from the States with a security clearance, but we had yet to be assigned one. Still, according to Rene, the local interpreter was the best in the country. He spoke several different languages, had served as a sergeant major in the Afghan Army, had seen combat, and knew everyone in the province.

The local team stood at almost full strength, with fourteen agents. They had a safe house which they had occupied for about six months. But there was an issue. For whatever reason, CSTC-A had yet to pay the rent on the place. Apparently, there was a problem somewhere at Campus Eggers with the paperwork.

Americans and their paperwork, I thought to myself. This was a phrase that I would use constantly with my Afghan friends, to explain why the U.S. military took so long to pay their bills.

The rent issue was becoming a problem. The property dealer (real estate agent) had started threatening the colonel in charge of the team. The actual homeowner, who lived in Peshawar, Pakistan, had been threatening both the property dealer and the colonel. I would need to address this immediately, before it got out of hand.

Rene explained I would have freedom of movement—but only if I operated under the radar. If it became public knowledge that I was leaving the prison camp in unarmored vehicles and without twenty soldiers to act as security, then I'd be shut down. Personally, I work better in small groups, without a lot of supervision. Operating in stealth-mode would be just my game. I decided to only let a select few military personnel know when I was leaving the base to conduct operations. Otherwise, I kept them in the dark so that they had plausible deniability in case I was ever killed or captured.

Rene offered to introduce me to all of the important players within the brigade chain of command, but I respectfully declined most of the introductions. The fewer people who knew my name and what I was doing, the better off I would be. Conversely, the more people who knew me, the more questions I would have to answer, and the more chance I'd have of being shut down by the battlespace renter.

The Fourth Infantry Division out of Fort Carson, Colorado, known as Task Force Mountain Warrior, oversaw Nangarhar, Kunar, Nuristan, and Laghman Provinces. Colonel G***** commanded that brigade, and maintained his headquarters aboard FOB Fenty. Lieutenant Colonel N***** served under him and ran Task Force Gryphon, which specifically oversaw Nangarhar Province. Lieutenant Colonel N kept her headquarters on FOB Finley-Shields, adjacent to where Rene lived.

FOB Fenty was also the home to a squadron of Apache Helicopters, a U.S. Army Special Forces element, the CIA, and Task Force 373. A whole cast of characters—or what military folks refer to as a "bunch of cats and dogs." Predator and Reaper drones also flew from the airfield.

***...I used to laugh whenever I passed by the Predator and the Reaper operations area.** Often, within this supposedly super-secret space, there would be a crew of Afghan day laborers cleaning or constructing something. Sometimes they would be within a shovel's reach of a Predator. I wanted to get a picture of the irony, but a ton of signs warned against it. People supposedly had their cameras confiscated for less.*

There may be strict secrecy on drone operations among the U.S. military and intelligence community, but when a missile hits a car in Pakistan, it's no secret. Even rural Pakistan has reporters, and it doesn't take a college degree to figure out who did it. Sort of reminds me of the "secret" bombing campaign in Cambodia during Vietnam. How do you bomb someone and keep it a secret? Once the bombs go off, I think it's apparent to the recipients (if they're still living), or at least their neighbors, that someone has bombed them. If you're in the tribal areas of Pakistan, that someone would be the U.S. government.

The other comical thing was that the Fox News website seemed to know exactly where the drones had hit and always reported on the attacks. I bet the

U.S. military and intelligence community turned to Fox News to see what the ground truth was. From their cubicles in Langley, it must be hard to determine who got killed in the drone strikes. A Predator can't dig through rubble and look at identification cards.

The next morning, Rene and I met for breakfast and he introduced me to my interpreter.

"This is Sammy."

"Nice to meet you, sir," Sammy said, as he shook my hand.

"Pleasure to meet you, my friend."

At the time, I didn't know how true the word *friend* would turn out to be. Sammy was about thirty-five years old, and he spoke excellent English, Pashtu, Dari, Urdu, Arabic, Hindi, and a couple of other off-the-wall dialects associated with the region. He stood about four feet eleven inches tall and couldn't weight more than a hundred pounds. I'm only five-foot-five, and I stood a full head taller than him. Sammy was dressed in traditional Afghan attire. (The Americans refer to this type of clothing as *man-jams*.) His appearance was that of a simple man.

At chow that morning, we chatted about the team. Sammy wasted no time in talking business. According to him, we had a lot of work to do. There were problems with the rent, problems repairing the vehicles, problems with the ammunition supply, and more importantly, the team had a plethora of information that needed to be disseminated. I liked Sammy immediately. He was obviously a worker and seemed to be dedicated to the cause.

I felt anxious to leave the prison camp and meet the team. But first, I needed to secure some Afghan clothes. According to Sammy, we controlled a shipping container full of humanitarian assistance (HA), such as bags of rice, beans, flour, sugar, clothing, blankets, and other basic articles. We could visit the container at the other side of the airfield and grab me a set of man-jams.

Sammy explained that an outgoing unit had given the container to the previous advisor on a drug deal. Due to internal bureaucracy and reporting requirements, it had become too much of a hassle for the military units to distribute the goods to the community. To hand out the aid, they would have had to build PowerPoint presentations showing the concept of operations (CONOPS), get levels of approval, and work through layers of coordination. It was just too much work to give stuff away to needy children. *Americans and their paperwork.* So, as the last unit departed, the previous ISU advisor gained *tactical control* over this container. In other words, a buddy of his gave him the key to it. Sammy explained that they used the HA supplies to pay informants, since most of the time they had no real money to offer them.

...The military folks on the prison camps never had any idea about what belonged to them, anyway. *I remember in Iraq, as they started closing down the camps and personnel started emptying out the shipping containers, they would open one and find a brand new pickup truck that had been in there for years or other much-needed items.*

Goods filled about half of this particular container, and a quarter of those goods had been bags of food. Rats ran rampant through the metal container and had already eaten through most of the bags of flour and rice. This killed me. People in the community were starving while the rats were getting fat. I blame American bureaucracy, paperwork, and inefficiency for that food sitting there, ruining.

In my spare time I run a small charity. Basically, friends and family members help me collect donated clothing and blankets. I take in small monetary donations as well, to cover shipping costs. Then, I box up the clothes and send them to contacts in Guatemala and Thailand for distribution to needy children. To me, the container was a treasure chest to continue my charity operations there in Afghanistan.

Sammy and I quickly located a box full of man-jams and I found a pair that fit. Coming from the HA container, they were of poor quality. Sammy suggested I have a set made so that I didn't look so poor. But poor was just what I was looking for. I didn't want to stand out in a crowd. I didn't want people to think I had money. I wanted people to look at me and think I was, at best, an automobile mechanic or a donkey-cart driver. I was going to get dirty, anyway. Why pay for a nice set of clothes?

Man-jams consist of a baggy pair of one-size-fits-all pants. Trust me. Three people could fit into those pants. For the single occupant, you had to weave a small rope through the interior to hold them on your ass. Sammy procured such a rope and he threaded it through the waistband with an ink pen. The shirt was a combination of shirt and dress, which hung down to my knees. A pair of sandals and a hat completed my Afghan uniform. (The Afghan sandals would end up hurting my feet so I resorted to wearing my shower shoes instead.)

I elected to wear the traditional Afghan wool hat, called a *pakol*. Ahmad Shah Massoud (the famous Afghan military leader who fought against the Russians and the Taliban) wore a *pakol*. He was a good guy and I figured I would follow in his footsteps. I also donned the twenty-dollar scarf I had procured from Campus Eggers. I thought the scarf went well with my man-jams.

After trying on my outfit, Sammy said that no one would think I was American—but he suggested that I get rid of the scarf. *What?* I paid twenty bucks for the scarf. I thought it looked good. Sammy agreed that maybe it looked good, but no Afghans in Nangarhar Province wore crap like that. Therefore, *Get rid of the scarf, sir. And by the way, you paid about nineteen dollars too much for that girly-looking scarf.*

I reluctantly agreed that I would only wear it on certain occasions. I paid twenty bucks for the damn thing and was determined to get some use out of it. When it got cold, I also planned to break out my fifty-dollar poncho and impress the masses. However, I wisely decided to keep the poncho a secret until the weather turned a bit cooler.

Now that I had the proper attire, I needed some gear. I took a cheap, blue backpack and stuffed it full of ammunition for my M-4 (rifle) and my M-9 (9mm Beretta Pistol). I also packed my GPS and a pile of batteries into the bag, along with unclassified maps of Nangarhar Province. A couple of flashlights, candy bars, crackers, cans of Coke, bottles of water, a roll of toilet paper, and a small first aid pouch completed my kit. I made sure all of my belongings stayed in that pack, in case we got hit and had to abandon the vehicle. It contained everything that I might need to escape and evade. I referred to it as my "ditch bag." Other people call them bug-out bags, but "ditch" better suited my situation. (The contents of the bag would eventually slim down to only ammunition and water.)

Now, I was ready to roll. We made plans to visit the safe house the next morning. That evening, Rene had the dubious chore of filling-in for the advisor who was on vacation. Rene spent hours answering e-mails and approving reports from other advisors embedded throughout the four provinces. I wasn't much help, since I had just gotten there and didn't really know what was going on. After working until midnight, Rene slept for about three hours before getting up and going to the gym. By seven in the morning, he was ready to go to chow.

I didn't have any interest in following his schedule. I had been working dope too long. I was used to staying out late, drinking until even later, and then sleeping in. I never eat breakfast, and you'll never catch me inside a gym. Rene was on his own with that routine. Still, at about 8:00 a.m., I was tooling around in the Forerunner, ready to roll out.

Sammy showed up a few minutes later.

"As-Salamu Alaykum, my friend."

"As-Salamu Alaykum, sir. How are you?"

This became the standard greeting during my time in Afghanistan, even with the gringos. I always try to build small words and phrases in order to blend in with the local people, no matter what country I'm in. It shows the average citizen that you're making an attempt at learning their language and customs. While saying *As-Salamu Alaykum*, it's customary in the Islamic community to put your right hand over your heart. Long after leaving Afghanistan, I still continue to put my hand over my heart when I greet someone.

A few minutes later, Rene joined us and we were ready to roll. We let Sammy drive, in case we got into a traffic accident. No need to complicate the situation by having a gringo at the wheel. We would raise attention to ourselves if that happened. (This mentality would quickly be altered, after a few mishaps with Sammy at the wheel.)

Our first stop took us to the far side of the airfield, where our shipping container stood in a secluded location. It made an excellent place to change clothes. Within minutes, we had transformed ourselves from two gringos and an Afghan to just three Afghans. I stuffed my M-9 in the waistband of my pants where it was well concealed but still easily accessible. Our M-4's could mark us as Americans, so we covered them up with blankets in the Forerunner in order to avoid any scrutiny at traffic checkpoints.

Our next stop took us to the initial exit point of the prison camp. As we rolled up, the Afghan guards instinctively stood in order to search the vehicle. If there were only Afghans in the vehicle, the guards were required to search it. As they began to open the back door of the Forerunner, Sammy broke the news to them that Rene and I were gringos. As Sammy spoke, both of their faces became puzzled as if they were listening to a drunkard talking nonsense. I think they thought Sammy was playing a joke on them. One of the guards approached the passenger window where I was sitting and said something to me in Pashtu.

"Sorry, I don't speak Pashtu, my friend. I'm American."

I showed him my common access card (CAC), which proved my identity. The guard looked temporarily stumped, but then broke out in a smile while yelling to the other guard. I'm sure his buddy was just as surprised, as they discussed our appearance.

The young U.S. soldier working at the checkpoint got up from inside the small hut to see what the problem was. I'm sure he assumed the Afghans were just chatting with their buddies until the guards waved him over. We held up our CAC cards for him to see and asked how he was doing. During our tour, we came to know this young soldier fairly well because he never left his post at the exit lane. He was also a true gentleman to the local Afghans, and they greatly respected him.

"Damn, I thought you guys were a bunch of locals holding up my line. Be safe."

With that, we passed through the initial exit point of the prison camp. The encounter had given us confidence that we could blend in with the public. If two locals and a fellow American didn't make us, then we were good to go.

At the second gate, which led out into traffic, we were held up for several minutes. According to Sammy, that was a good sign. If the guards at the gate knew that Americans were riding in the car, they would have stopped traffic to let us off the base quickly. But, if they only saw Afghans in the vehicle, they would wait until it was convenient for them to open the gate. As we turned left out of the entry control point (ECP) and headed toward the city, I didn't feel the strangeness that I had felt my first time in Iraq. I felt like I was right at home among the Afghans.

On our way to the safe house, we passed the entrance to FOB Finley-Shields and Camp Hughie. Rene explained that the Afghan National Army (ANA) stationed there were responsible for the camp's security, and they were obviously more militant than the contracted security guards who manned the exterior of FOB Fenty. If we had to go to Camp Hughie at night, he told us, be especially careful when approaching those guards.

"No problem," replied Sammy to Rene's line of caution. That was a phrase I would hear thousands of times during my tour. In Afghanistan, it could mean that a car had a flat tire, and it was *no problem* to fix it; close to reality. Or, it could mean that you were surrounded by hundreds of bad guys and it was *no problem* to kill them all with a butter knife; far from reality. Through experience, I mainly came to interpret the phrase *no problem* to mean that a situation was seriously screwed up, and needed immediate intervention to prevent further chaos.

We snaked our way through the City of Jalalabad, past the "Big Mosque" and park, the bazaar, and the Provincial Police Headquarters (PHQ) compound. Then we turned onto a dirt street pocked with huge potholes. Open sewers, common to this part of the world, gaped on either side of the street. As we bumped along, we watched young children playing in the sewage. Sammy pulled the Forerunner up to a large blue gate in front of one of the houses and honked the horn several times. While I thought this was a bit rude and brought attention to us, I soon learned that Afghans see their horns as a tool and using them is the accepted norm.

A small slit that had been cut into the gate abruptly opened. We saw a set of eyes peering out at us, before it quickly slid shut again. It reminded me of those old movies with the speakeasies from back in the twenties. The doorman opened the gates and we drove into the qalat.

In Afghanistan, most homes are built as qalats. This basically means that a high wall protects the entire compound, and the home sits somewhere in the middle, like a small castle. Qalats typically have only one entry point, the front gates, thus making them easy to defend.

The walls around the safe house were adorned with concertina wire, and thanks to their height, provided the utmost privacy. The safe house had a nice concrete parking area in front of the home, and even had two covered parking bays. While the home itself appeared a bit older, it was a palace by typical Afghan standards. As we exited the vehicle, a group of Afghans stood up from a table on the front porch and came to greet us. Sammy met each one with a hug and handshake, and then he began the introductions.

"This is Mr. Mark and this is Mr. Rene."

Sammy introduced us to each of the fourteen team members by name. I didn't remember a single one at the time because I'm terrible with names, but since these were going to be my boys, I made it a point to learn them as soon as possible.

"And this is the colonel," Sammy said, as he gestured to a middle-aged man in his late forties to early fifties.

In Afghanistan, it is very difficult to guess a person's age and one should use caution before rendering judgment. The years of war and hardship take a toll on a body. A person that's twenty-five could easily look forty-five depending on how his life has played out.

The advisor who was on vacation had called to warn me about the colonel. Apparently, he and the advisor who had just left had planned to recommend replacing the colonel. They didn't feel he was strong enough to do the job. It wasn't that he was a bad person. It was just that he had been a communications officer prior to taking on the role as site leader. Being in communications doesn't take the same balls as leading a high-risk team of operators. I decided to give him the benefit of the doubt and conduct my own assessment. I would end up being thankful that I did.

Sammy and the colonel gave us a tour of the safe house. As we entered the two-story home, we first passed a room that obviously served as the hangout area. Twin-sized mattresses lined the floor against the walls, and a big screen LCD television stood in the corner.

Across the hall from the hangout room stood a makeshift conference area. Someone had pushed together several small desks to make one long conference table, and about twenty chairs clustered around it. Various maps clung to the walls, showing different views of Nangarhar Province. Obviously, the previous advisor had donated these because they were all in English. As I inspected the maps, I realized that they were unclassified and didn't even show the military grid reference system (MGRS). Instead, they were marked with latitude and longitude. Nobody uses latitude and longitude. Everyone uses MGRS—which is not classified. In fact, most commercially available GPS units can display MGRS.

I laughed to myself that we would give them these useless maps. I eventually realized that the maps marked with MGRS were classified as limited distribution, or LIMDIS. This made personnel afraid to give them to the Afghans because it might be a security violation. Of course, the Afghans had lived in Afghanistan for their entire lives. They know the terrain much better than the Americans do. I think it probably would have been OK to give them proper maps.

...And another comical thing. *If a map showed the locations of the prison camps, then it had to be classified as "secret." Now, I could understand if a base was hidden and, therefore, "secret," but Afghans worked at the prison camps, delivered goods, and provided services there. Their locations weren't secret. Plus, Afghans could drive past the camps and use their GPS devices to find the MGRS coordinates by themselves. But, if you gave an Afghan a map with the locations of the camps marked on it, then you would go to jail. So, during my tour, I couldn't carry any of those maps. Instead, I had to constantly ask, "what camp is near there?" in order to report information to the right people.*

During our tour of the safe house, the colonel summoned one of the agents who served as the logistics officer, and he showed off the equipment room. Behind its locked door were shelves of handheld radios, rifles, GPS units, first aid kits, and a plethora of other supplies. This was good news, and I intended to build some inventory spreadsheets as soon as possible. Technically, once CSTC-A issued the equipment to the Afghans, we Americans no longer had responsibly for it. The Afghan government became the property owner and if anything was lost, they had to deal with it. However, I wanted to have a clear picture of what supplies we had and what we needed. I wanted to find out if anything shady was going on, as well.

The colonel ordered the logistics officer to show us his property book. He opened it, revealing several pages of flowing Dari. Of course, I had no idea what it said, but the pages looked neat and orderly. Obviously, he kept a good accounting and the colonel was proud of that fact.

I was duly impressed. I had heard so many negative things about the Afghans and corruption, I just assumed that I'd see fraud like I'd witnessed in Iraq. But the Afghans definitely didn't have anything in common with the Iraqis.

...*In Iraq, for example, the U.S. purchased brand new Glock pistols for the police officers.* *As a taxpayer, this in itself pissed me off. Why didn't we give them the old Beretta 9mm pistols that the military had been carrying for years? Then, we could have issued the brand new Glocks to our soldiers. Plus, in my opinion, the Glock is more reliable than the Beretta and doesn't require as much cleaning or maintenance. What's worse, after issuing the Glocks to the Iraqis, most of them were promptly "lost." In other words, those thieving idiots sold them down at the market and pocketed the money. They would show up for their shifts with no pistols and no explanations. This was your tax dollars at work, and we were negligent enough to let them get away with it.*

The rudimentary kitchen wasn't clean enough for my standards, but it seemed to serve its purpose. Sammy explained that the agents cooked lunch each day and ate together as a team. One of the guys was a great cook. We passed by the kitchen and turned toward the stairway, which was blocked by an iron door and bars. Sammy explained that the house had been used as a polling station prior to the agents moving in. The previous tenants had installed the gates in order to lock down and secure the second floor.

The upstairs revealed a room that the logistics officer and the admin officer used. Their laptops stood open on their desks and a file cabinet sat in the corner. Every minute, my perception of the Afghans was changing. I had read so many articles about the low literacy rate in Afghanistan, but these guys were generating reports using Microsoft Word and Excel. In fact, they made it a point to tell me they desperately needed new printer cartridges and paper.

We ended up in the colonel's office. In their culture, the top dog enjoys the best of the best, without proportions. (Actually, that's much like U.S. government employees, too.) A huge mahogany desk dominated his office, along with a big screen LCD television, leather sofa and loveseat, and a nice coffee table. An assortment of plush Afghan rugs covered the entire floor. As we entered, the colonel turned on the air conditioner with a wireless remote control.

A young ANA soldier entered the room and poured tea. I later learned that he was the colonel's nephew. The colonel had made a drug deal with the boy's commander, to get him assigned to the team as security for the safe house. Not a bad gig for the kid. While his comrades were out in the field, humping mountains with a hundred pounds of gear, he was pouring tea and wearing man-jams.

In Afghanistan, everything starts with a round of tea. The tea is usually accompanied by a glass plate of nuts, raisins, and candy. Afghans extend all this as hospitality to visitors. I would soon learn that if you arrived anywhere close to mealtime, then as a further courtesy, they would invite you to eat with them. It was disrespectful if you declined. During my time in the country, there were many days that in our haste to get things accomplished, we would arrive at someone's office too close to lunch and have to eat with them. Sammy and I often wound up having to eat twice, if we had scheduled multiple meetings too close to noon.

Through Sammy's translations, the colonel expressed his welcome and thanked us for coming to Afghanistan to help the people. The colonel acknowledged our sacrifice of being away from our families. After the greetings and formalities, he turned to the business at hand.

The first order of business involved the safe house rent. The team had occupied the safe house for over six months, but the property dealer hadn't seen a dime. This had become a huge problem for the colonel. When CSTC-A signed the lease agreement, the property dealer had asked the colonel to give his personal guarantee. This is the way business works in Afghanistan. The Afghans don't use contracts and a ton of paperwork. Instead, they use what used to suffice in America: a handshake and a personal guarantee. While the actual contract was technically between the homeowner and the U.S. military, the property dealer had insured the deal by obtaining the colonel's personal guarantee. If CSTC-A didn't pay, the colonel would have to come up with the money.

Instinctively, and although I didn't know it at the time, very naïvely, I assured the colonel that I would talk with the contracting folks and work out the problem. *This should be an easy fix*, I thought. I would soon learn how wrong I was.

Another problem, the colonel explained, involved fuel. They couldn't get diesel fuel on a regular basis for the generator. They had requested it through Afghan logistics channels in Kabul, but hadn't received any. Fortunately, a small amount had come through CSTC-A channels, and they conserved it by only running the generator for a few hours a day. I inquired and they informed me that the house wasn't connected to the city's limited power grid. They explained that in order to get it connected, you'd first have to bribe a few government officials. They could do it, but who would provide the funds? Therefore, the team had only limited electricity supplied by the generator.

They also had no way to fuel their vehicles. No gasoline had come through from either Afghan channels, or via the CSTC-A route. The previous advisor had been taking them to FOB Fenty and getting gasoline from the Americans, but since he had departed they no longer had access to the FOB.

There was no maintenance contract in place for the generator. The colonel estimated that the generator had cost CSTC-A over thirty thousand dollars, and another three thousand to have it installed. It needed to have its filters changed on a regular basis to avoid any mechanical problems. It would be a shame to ruin a brand new generator over something as small as changing the filters, according to the colonel.

Next, the vehicles desperately needed repairs. Anytime one of their cars or motorcycles broke down, he and the agents had to pool together their own money to fix it. The colonel opened up a desk drawer and pulled out a stack of various receipts written in Dari. He explained that they kept meticulous records on the money spent repairing the vehicles, in hopes that someone would eventually reimburse them. The colonel gave them to me, if not more than to pass the responsibility. Now, he could tell the men he had spoken to the American about it and passed on the receipts. It was false hope, but hope nonetheless.

We discussed a flurry of other administrative issues that day, from agents not receiving their promotions, to some of them being owed back pay. The colonel told me confidently that I needed to call "The Jims" in Kabul to work out these problems. I'm not sure how it started, but everyone knew Jim as "The Jims." Later, I came to realize that they weren't saying the word "the" before "Jims." They must have thought his name was *Tha-jims*.

I was a bit overwhelmed with all the administrative issues. I've never been one to endeavor to be the boss. To me, the boss of any organization has to deal with too much paperwork, too many meetings, and too much bean counting. I always made sure to avoid any potential for promotion. I like being an agent and I like being an operator. I like the idea of only being responsible for myself, chasing dope, and catching bad guys. The military had hired me to act as a subject matter expert on how to conduct operations—not as an expert on how to get fuel for a generator. I really wanted to discuss the team's ongoing operations.

"Colonel, what kind of cases are the agents working on right now?"

An exchange of words between Sammy and the colonel ensued.

"Ah sir, he says they have a meeting every morning in which the agents give him a report from their informants."

The colonel pulled out a small notebook from the inside pocket of the vest he wore over his man-jams. The notebook measured about four inches by three inches, way too small to take detailed notes in.

The two exchanged words again, and then Sammy gave the rundown on about five different Taliban groups operating within Nangarhar Province. They knew their names, villages, and areas of operation. It was all very good information. I copied down what Sammy reported and planned to disseminate it among the intelligence community.

After that exchange, I asked Sammy to tell the colonel we had to get back to the prison camp, because Rene had a lot of computer work to catch up on. I promised to inquire with "The Jims" about all of the administrative problems, and told him that we'd return first thing in the morning to attend his 8:00 a.m. meeting. I wanted to address the team at that time and formally introduce myself. More so, I wanted to see how the colonel conducted business and what everyone did throughout the day. I also wanted the team to know from the start, that I was there to work. After a round of handshakes and hugs, we departed. It had been a good day for me.

● ● ●

The next day, true to my word, Sammy and I rolled into the safe house at 6:00 a.m. sharp. Notoriously, Afghans start their day early. By custom, they arise early to pray, but it's also just part of their culture to get up before sunrise. Even so, we beat the colonel and most of the agents to the safe house that day, which was our objective. I needed to set the precedent that I led by example. I also wanted to see who arrived early, who got there on time, who showed up late, and who failed to show at all. The generator wasn't running when we arrived, and a bluish, predawn darkness hung in the courtyard of the qalat. Sammy and I sat down at the table on the front porch and sipped some tea with the duty agents and the young soldier. They anxiously explained that the colonel was on his way and would get there shortly. Apparently, they had called his cell phone upon our arrival, to let him know that the new American had already returned to the safe house. I'm sure the poor colonel felt he had to break his morning routine and hurry to meet us. He was probably thinking, *Why in the hell is that American already there?*

The colonel arrived about forty-five minutes later, and immediately began offering excuses as to why he was late. I asked Sammy to tell him not to worry. As the colonel, he set the work hours. I also explained that we had shown up early, because I would rather spend time with the team, versus being cooped up like a sardine on the U.S. prison camp. The colonel laughed and seemed a bit more at ease.

We all assembled in the meeting room for the 8:00 a.m. gathering. The colonel presided from the head of the makeshift conference table, and Sammy and I sat to his immediate left. As he spoke, Sammy translated his words in a low voice. After the colonel had his say, he then directed each agent to give their report, while he took notes in his four inch by three inch notebook. The agents recounted detailed information about Taliban commanders, their locations, how many fighters were with them, and what kind of weapons they carried. It was definitely good information and Sammy wrote it all of down for me.

Toward the end of the meeting, the colonel formally introduced me to the team. I thanked him and began speaking directly to the men as Sammy translated. I have learned that when speaking through an interpreter, it's best to keep the conversation simple and direct, and to pause frequently for translation. If you don't pause often enough, the interpreter will forget what you've said and lose parts of the conversation. All the pauses make for a time consuming dialog, but it's important to get the proper message across.

…The American military teaches personnel to use an interpreter like a tool. The interpreter is there only to translate what you say. The interpreter should sit behind you, so that the focus is on you. Keep your interpreter in check, and so on, and so on. That's bullshit. If you want to interact successfully with another culture that you know nothing about, you had better build some trust with your interpreter and give him the leeway to

assist with the conversation. It's the message that needs to be conveyed. Not the step-by-step procedure or who-the-hell's in charge. Your interpreter will keep you out of trouble if you let him. If you don't show any respect for him, you can bet that he'll translate what you've said, word for word. Afterward, he will laugh his ass off when you've made a fool of yourself and lost credibility.

Through Sammy, I explained that I had previously served as a narcotics agent in the States, and had much of the same experience they did. I wasn't there to tell them what to do. I was there to share some of my training and lessons learned with them, to deal with the Americans and their paperwork, and to eventually walk away from a self-sustaining team. I considered them my brothers and would never ask them to do something that I wouldn't do myself. I realized that the team had problems. Anytime you start a new unit, there will be logistical problems. We would work through them. The only thing that I requested was that everyone report to work motivated about doing their jobs. Our efforts would save Afghan lives, including women and children—possibly even some of their relatives. Roadside bombs show no discretion. We had the authority and responsibility to stop the people building and planting the bombs. Tax dollars from the United States and from the Afghan public paid our salaries. They paid us to do the dirty work, and they were counting on us to perform.

While I didn't want to come across as the "New Sheriff in Town," I wanted them to know that I was serious about the work. I explained that serving on the team put them in a position of high-risk. If anyone felt that the job wasn't for them, I would gladly help them obtain a transfer to a much less dangerous line of work. There would be no shame in it and no hard feelings. Just let me know. I was serious about that, too. I would have helped any of them find an easier job. But saying so served another purpose; it let them know I would hold them accountable for their performance.

Throughout my spiel, all eyes stayed fixed on me without even the hint of anyone catching some Z's. I think that Sammy and I may have even inspired a few. They seemed ready to do some real work.

● ● ●

Back on the prison camp, I put the finishing touches on a ten-page spreadsheet. It itemized all of the equipment "The Jims" had documented as having been issued to the team. I asked Sammy to call the colonel, and tell him to have the agents bring all of their issued gear the next day for an inspection. That probably went over like a turd in a punch bowl because of the short notice, but I needed to complete this quickly. How could I ask the gringos for more equipment and supplies, if I didn't even know what we already had on hand?

The next morning, a full house was waiting as Sammy and I pulled through the gates. They offered us the standard greetings as well as some tea, but I declined. I was a man on a mission. It wasn't a culturally correct move, but by the end of the day I wanted those spreadsheets complete.

We inventoried the motorcycles, vehicles, weapons, computers, printers, communication gear, and anything else that had a serial number. By the end of the day, I had accounted for every piece of equipment. I couldn't believe it. Everything was there. The colonel and the logistical officer were all smiles. They knew they were squared away and were proud as hell to show me.

I asked them to empty the magazines to their Makarovs and show me how many rounds they had. Each agent's total count varied from between six to ten. That was all the ammo they had for their sidearms, which their lives might depend upon. It was totally unacceptable. I felt responsible and realized that I'd have to find some ammunition to plus them up. I felt guilty that we'd set up a team without properly arming them. Someone should have been fired for this poor planning. With all of the weapons, ammunition, and money the U.S. had sent to Afghanistan, this was the best we could do? Ask any cop in America about how important their handgun is to them, and listen to their response.

Erroneously, I didn't ask them to empty their rifle magazines. I merely went down the line and visually inspected them. I irresponsibly made the assumption that the magazines were full. They used AK-47 ammo, which obviously wasn't hard to find in Afghanistan. Later on, during a pre-raid weapons check, I picked up an agent's magazine. It felt a bit light, so I shaved off the rounds. The count was ten. I discovered that most of their magazines had less than ten rounds. They should have had thirty rounds each. I later apologized to them for not doing my job. Once I realized the mistake, I immediately went about acquiring ammo for the rifles. Within a couple of days, I had scored a whole case via a drug deal with some good-hearted civilians at the ammo supply point. Thanks to those sorts of actions, the agents came to respect my inspections and pre-operational checks, and they realized the checks weren't design to get anyone in trouble. I did them so that I could identify any deficiencies in their gear and then try to get them what they needed. The inspections weren't performance evaluations; they were merely readiness drills to ensure their safety.

By the end of the day, I returned to the prison camp and began entering data into my trusty spreadsheet. Now, if "The Jims" wanted or needed anything, I could send him the data in one complete document. It would save me time in the future, whenever someone wanted to count beans, generate paperwork, or conduct inventories.

● ● ●

I stuck to my plan of completing an evaluation period of the team before I made any changes. I observed and learned for almost two months. I got to know the agents and learn their personalities, and they got to know me. I wanted them to look at me as a trusted friend, instead of some American they had to listen to for a few hours each day.

During the first couple of months, I got the agents some training on surveillance and photography from contacts at the prison camp. These contacts were also cops from the States. They weren't being utilized properly, according to their contract, and sat around idle most of the time. They were eager to help out and do some work.

I convinced the colonel to divide the agents up into three, four-man teams. It worked better that way due to our vehicle situation and for command and control. We merely referred to the teams as *One*, *Two*, and *Three*. Simplicity. This sat well with the men, especially after the last set of team names the Americans had inflicted on them.

The Americans who created the unit had originally split them into two teams, called the Spider Team and the Cobra Team. Obviously a gringo came up with those names. One day early on, while at the safe house, I asked the agents which of them were on the Cobra Team. Several raised their hands. Then I asked who was on the Spider Team. The rest of the men looked around with shameful expressions and then reluctantly raised their hands. *Did I say something wrong?* Nobody wanted to tell me and I asked Sammy about it. He explained that they didn't want to hurt anyone's feelings, but the agents felt embarrassed to be on the "Spider" Team. Apparently, in Dari (or it could have been Pashtu), the word for spider is *jolaa,* which also loosely translates to mean a "dressmaker." It implies that someone is weak and soft. Basically, it means you're a *pussy.* It would be the equivalent of naming the Detroit Lions, the Detroit Kittens. Not very motivating, and it doesn't build camaraderie or team pride.

That's typical of us Americans. We make choices based on what we like, without considering the locals' perspective. We then force them to do things our way. After I learned about the issue, I assured the agents they'd never be referred to as Spider or Cobra Teams again. We would simply assign the agents to Team One, Two, or Three. That assurance brought smiles of relief. It would also eventually create a three-way competition at productivity.

● ● ●

We slowly began to run cases. At first, we went after various pieces of ordnance or small weapons caches our informants had stumbled across. In Afghanistan, there's so much ordnance left over from the Russians, not to mention what was supplied by the Americans, that the country will never be completely cleared. Going after the small-time ordnance wasn't exactly complicated investigative work, but at the time the team was in the crawl phase. By taking these cases, their readiness began to improve. They got the chance to practice their coordination and operational planning skills, and identify any problems before we took on a real operation.

One of these early cases was simply an old piece of unexploded ordnance (UXO) left over from Russian times. An informant had found it near an area where children gathered to play. As soon as he contacted one of the agents, one of our teams responded to the area. They confirmed its identity as a Russian mine and established a perimeter around it. I called back to the prison camp and asked the duty officer at Task Force Paladin to have an Explosive Ordnance Disposal (EOD) unit sent out. We were close to the prison camp and I didn't think it would take them very long to arrive—but then I started to learn about reality in Nangarhar Province with the U.S. military and Task Force Gryphon.

While Task Force Paladin had EOD technicians with Mine Resistant Ambush Protected (MRAP) vehicles at the compound ready to roll, this area was not their battlespace. They weren't authorized to respond to a request for assistance, even if it came from a fellow Task Force Paladin member. The battlespace belonged to Lieutenant Colonel N, who resided on FOB Finley-Shields. Once a request for EOD was made and routed to her operations center, she could decide whether or not to dispatch an EOD unit. *What the hell do you mean, they may or may not respond?* As a law enforcement officer, this notion was foreign to me, but I quickly learned they had that luxury. Unlike police officers who have no choice but to respond to a situation, no matter how dangerous, the U.S. military can elect to stay within the safety of their prison camps. They can just say screw the Afghans, and to hell with the safety of the local people. Apparently, Lieutenant Colonel N was more concerned with her fitness report. Rumor had it, she had issued a personal guarantee to all of her soldiers, promising to bring them home safe from Afghanistan. I'm not sure how she could, even in good conscience, make that promise, but that was the chow-hall rumor. Over time, I started to believe it because she wouldn't allow any of her people to leave the camp if the slightest risk existed. Meanwhile, the bad guys ran unchecked in Nangarhar. Her inaction was creating a dangerous situation for whoever would take over for her in a year.

After getting frustrated with the gringos and their lack of enthusiasm about assisting us, I quit calling. That's when Sammy told me I needed to call Ralph.

Ralph is an old, retired U.S. Navy master chief and EOD tech. The U.S. had brought him to Afghanistan to establish and train an Afghan team to deal with unexploded ordnance. Ralph's unit, called the Weapons Reduction and Abatement team, or WRA for short, operated out of a rented house in Jalalabad, much like we did. Local Afghans made up the team, and they rolled in soft-skinned vehicles with a few contracted Nepalese security guards. Ralph was basically my counterpart in a different field. He ran a reactive EOD team, while I ran a proactive counter-terrorism and counter-IED team. We found the IEDs, UXO, and bomb makers, and then Ralph's team did the dangerous work of dealing with anything that went boom. Our teams developed a wonderful relationship, especially since the U.S. military never seemed motivated to assist with our efforts.

I called Ralph and within ten minutes, he and his team were on the scene. Ralph's crew safely retrieved the ordnance and secured it in their vehicle. Later, they took it to their storage lot and eventually destroyed it. Problem solved.

That night, I returned to the prison camp pissed off at my own people for letting me down. Nothing was said and no excuses were offered, other than the fact the battlespace renter wouldn't allow the EOD unit to roll out. It bothered me, in particular, because U.S. EOD technicians are the best in the world. They endure far more training than any of their counterparts in other countries. They have the latest and greatest equipment such as bomb suits, robots, telescopic video cameras, and they are proficient in the use of explosives. Plus, they roll with a security detail of at least four MRAPs and twenty young grunts armed to the teeth. The U.S. taxpayers spend a lot of money on EOD techs and their equipment. Millions of dollars in training and equipment sat idle that day because an incompetent battlespace renter wouldn't let them do their jobs. Apparently, Lieutenant Colonel N thought it was too dangerous to allow her people to roll about a mile away from their gates.

...Let me say something about my EOD friends: *They were always ready to come to our assistance. They are true warriors and are scared of nothing. Matter of fact, they were sitting in their trucks that day, chomping at the bit, ready to roll. They just couldn't get approval from Lieutenant Colonel N to depart friendly lines. I guess LTC N didn't think that a mine lying near an area where Afghan children play, warranted a response. But I have to wonder; would she have dispatched an EOD unit if there was a land mine near American children, at a playground in Arlington County, Virginia?*

Although Ralph's crew was competent, they didn't have the robots, bomb suits, or other gadgets that the Americans had. I have come to believe that the Americans just need to give the Afghans their vehicles and equipment so they can do their jobs more safely. What good is the equipment if it's just going to be locked down at one of the prison camps?

● ● ●

A few days later, another informant called to say he had located four RPG rounds. No big case at all, but an RPG round will cut through the side of an MRAP like a hot knife through butter. We traded him some rice and beans for the rounds, which were in pristine condition and looked as if they had just come from the factory.

This particular informant had an interesting angle. Come to find out, he and a small crew of men and donkeys were raiding Taliban ammunition caches at night. They were thieves—but with a noble cause. At one o'clock in the morning, while everyone else slept, the raiders and their sleepy donkeys were humping ordnance out of the danger zone and into safe havens for pickup. They had balls.

The informant, who I nicknamed "The Cleaner" on account of his tactics, advised me that he was working on something a little bigger. According to his information, a friend's family had two Stinger missiles still in their cases. He reported that they had built the Stinger cases into a mud wall inside their house, and that the missiles had stayed hidden there for years.

At first, I threw the bullshit flag on this one. It sounded too good to be true. We had received leads on Stingers before, but they never panned out. We always found Milan anti-tank missiles instead. Over the years, the Americans had bought back or seized several Stingers, and as the years ticked by, they had become scarce. If anybody had one, logic says they would have gotten rid of it and made big money by now. So, why in the hell hadn't these folks already dug them out of the wall and sold them? They'd be living high on the hog somewhere with a pocket full of money.

"You guys recover caches all the time. Why didn't those people turn them in years ago?" the snitch asked. He had a good point. He went on, "Some people are scared to turn them in, or they're scared they'll get arrested. They don't trust the Americans." Another good point made by the snitch.

We put the man on the job and asked him to focus on getting the Stingers. His friend's house stood only a few kilometers from the Pakistan border, in the northern most part of Nangarhar Province. The CI planned to meet with the friend and convince him to sell the Stingers to the Americans. Everyone knew that the Americans paid a lot of money for Stingers, and at the time, they were worth a hundred grand a piece.

If he could convince the friend, they would transport the Stingers via donkey, to the nearest FOB. We would be standing by at the FOB to coordinate with the gringos, and call off any American air power that might be tempted to target them while in transit. We would receive the Stingers, and later on, the snitch would travel to Jalalabad to get paid.

If the family would not agree to sell the Stingers, however, the informant explained that his crew would resort to alternative measures. Being a law enforcement agent, I really didn't want to hear his sinister plan. I'm sure they planned to roll in strong and rob the place, but that wasn't my problem. Besides, if the U.S. military thought there were Stingers in the house, they would probably drop a couple of bombs on it and kill everyone inside. So what's the big deal about a robbery, under those circumstances? Getting robbed beats getting dead from U.S. air power. We told him we'd be ready when he called.

CHAPTER 3

Come to Jesus

After observing the team for several weeks, I had made a number of observations. First, the colonel was not a strong leader. He wasn't the type of guy who could order someone to his death in order to accomplish a mission. He was a bit soft, and the men took advantage of him because they knew he wouldn't get on their ass about things. However, he was meticulous about paperwork and accountability. He excelled at all of the tedious tasks that went with his position. I would never have to worry about whether or not he had filed the appropriate paperwork in Kabul, completed a report, or accomplished administrative duties. He was thorough with those types of issues.

As far as I was concerned, he was exactly the guy I wanted in charge. I wanted no part of the administrative work. I would handle the dirty work, and he could push paper. There's a saying in Afghanistan: *Someone asked the bullet why he was moving so fast, and the bullet responded by saying, it was because of the powder behind him.* Sammy and I would be that powder.

Second, I realized that the team members didn't understand what the hell they were supposed to be doing. They thought their jobs were to talk with a few informants every day and then report the information to the colonel the next morning. They also mistakenly thought that no one could know who they were. Because of this belief, they kept a low profile and didn't cooperate with other law enforcement entities in the province.

This made them, in a sense, just another one of the hundred or so intelligence entities operating in the country. They had turned into another report-writing unit that made absolutely no difference in the fight. Writing reports and taking down information doesn't have an effect, unless you go and do something about it. In America, we know who all the drug dealers are. That's the easy part. The hard part is putting them in jail and keeping them there. None of the three-letter intelligence agencies cared about that aspect. As long as they could put some information on paper, they were doing their job. It didn't matter if the information was true or not. They were only there to write reports. That's how they were evaluated. My team would not be report writers. They were law enforcement agents and would start acting as such.

I had identified the problems and was ready to start fixing them. I sat down with Sammy and explained the plan of action. I needed to become the "New Sheriff in Town," or else we would never accomplish our mission. After Sammy understood what I meant, he seemed reluctant to take the hard-line approach.

"I don't care, Sammy. I realize you're tight with the guys, but they're not doing what the hell they're supposed to be doing. It's time to change their mentality, immediately."

As Sammy and I departed the gate of the prison camp, I began to go over our strategy for the pending all hands meeting.

"This is going to be a Come to Jesus meeting with the troops."

Sammy just looked at me. Without him having to say anything, I realized my mistake. I'm bad about using inappropriate American sayings around my Muslim brothers. On another occasion, I once told the colonel that we didn't want to be *the last hog at the trough* when it came to picking up new equipment. Islam, Jesus, and pork don't mix.

On the way to the safe house, we mentally prepared to get gangster on the team members. I use the word "gangster," because it's often the way you have to deal with players in Afghanistan. It's all they know. You either have to be the biggest gangster in the room or they will take advantage of you. The guy in charge has to carry the biggest stick and not hesitate to use it. After years of tribal infighting, invasions, and general war, it was the way of the land. You didn't earn the name "warlord" for being nice, but nonetheless the "warlord" was in charge.

...American policy tells everyone to *be politically correct and sensitive to the host nation's leadership. In other words, appease them and don't worry about whether or not you accomplish anything. Don't worry about whether or not you hold them accountable for the waste or misuse of U.S. tax dollars. Just be politically correct and don't offend any of their customs or beliefs. (Even though that was the official guidance, Americans continually subjected the Afghans to unnecessary searches and indignities, and showed no real respect toward them, anyway.) That advice is actually relevant for the majority, who* *show no interest in developing real relationships and friendships with the locals. If you develop a true friendship, you don't have to worry about political correctness and can speak ground truth. You can address difficult situations without the worry of offending someone.*

In any private company, no matter your religion or beliefs, if you don't do a good job, you will no longer be employed. This sort of accountability obviously runs counter to the way the U.S. government operates. If you work for the U.S. government, you just have to show up and you'll get paid. Government workers are not evaluated or paid based on their performance or productivity—in case you haven't noticed. Our team wouldn't operate with the U.S. government mentality.

My "Come to Jesus" meeting didn't have anything to do with religion or culture. There would be no political correctness. We had a job to do, but we weren't doing it. Business is business.

As the team assembled in the conference room, I could sense an air of anxiety. Maybe Sammy had dropped them a subtle hint that the meeting wouldn't be pleasant. Occasionally, I would insist that Sammy flip over to strict translation mode. In other words, don't paraphrase what I've got to say, don't quarterback it, and don't change the message. This would be one of those times. Through Sammy, I began the proceedings.

"Gentlemen, I first want to ask all of you a question. What have we accomplished in the past two weeks?"

As everyone looked around at one another, obviously confused as to the question, the colonel pulled out his trusty little notebook. He began to read off excerpts from all of the reports they had provided, while he dramatically flipped the pages. I stopped him, in order to tell a story. It was a long story but I had all day. Since we had no cases developing, it wasn't like they needed to be anywhere, either. With a captive audience, I began to tell the tale.

When I was assigned to the Drug Enforcement Administration (DEA) Task Force in Atlanta, Georgia, I had a group supervisor named Mel. Mel passed away several years ago; rest his soul. Mel did a couple of tours in Vietnam as a young Marine, before embarking on a thirty-year career in narcotics. He worked for DEA from its inception and had an adventurous career fighting the *Drug War*. As Mel was close to retirement when I worked for him, he really wasn't concerned about policy or politics. That was the best kind of person to work for. He was *old school*.

Occasionally, Mel would come out of his office and survey the group of thirteen agents steadily typing away on our computers. He would tell everyone to stop what they were doing and come in front of his office for a meeting. As we all gathered around the old couch and wooden table, the questioning began.

"What have you done the past week?" Mel would ask, as he pointed to the first unlucky agent.

"Well, boss, I've written four reports, went to court..."

"Shut the hell up! You haven't done shit!"

The next unlucky agent suffered the same outcome, and so on.

"I've come out of my office every day this week and looked at this table. I haven't seen any dope, guns, or money on my table. I haven't seen one shithead being interviewed in my interview room. Therefore, you all haven't done anything. You've been sitting in the office. What the hell are you waiting on? When was the last time a drug dealer came to the federal building, asked to talk to one of you, and then turned over his drugs, money, and guns? Never! The bad guys aren't in the federal building. So, what I want you to do, is get the hell out of my office and not come back until you've got something to put on my table. I don't care how long it takes you."

At that point, he ordered us to pack our bags and vacate the office. As we all dragged ourselves down to the parking garage, a huddle would ensue. The mood was initially comical at being thrown out but quickly turned serious.

"You got anyone we can go lock up?"

"No. What about you?"

"I've got a good location on one of our low-level targets. If we can talk our way into his crib, he should be good for some guns and maybe a kilo or two."

"Let's go. I've got to get back into the office to finish a report the AUSA (assistant United States attorney) has been riding me about."

"Me too, I'm knee deep in a wire affidavit that's going to be stale if I don't get it submitted."

After coming together under stress and the proper motivation from Mel, we would quickly have someone in handcuffs and some dope to put on Mel's table. When he walked back into the office, we would have everything displayed in the most dramatic manner, just praying that it met his threshold of accomplishing something.

"Now that's what I'm talking about, where'd you get him from?"

With that, we were no longer *persona non grata* from our cubicles. Life could return to normal. We could continue typing away, satisfying the internal reporting requirements of the federal government. We could continue the non-tangible work which often becomes the norm within federal law enforcement. Unfortunately for the taxpayer, the federal government doesn't care about producing tangible results. As long as you can generate enough paper and kill enough trees, you are perceived to be a good employee. Don't worry about actually doing something.

There is a saying among federal law enforcement agents that summarizes the mentality: "Big cases? Big problems. Small cases? Small problems. No cases? No problems."

After explaining this story to the team, I told them their days of merely writing down information were over. Sure, the information was good. But, if you wrote down some information, I was going to want to know what we were going to do about it. Writing down information was not producing tangible results. That was not "putting dope on the table." If you couldn't physically put something on the table, it didn't count. If I couldn't touch it, take a picture of it, or talk to it, you had done nothing. This was the new standard. We would adopt the mentality of having big cases, even though it meant having big problems.

If anyone didn't like it, I would, with much remorse, kick him off the team. We had a duty to protect the Afghan people. At the time, I didn't view them as having that sense of responsibility to the people. They didn't understand the words "civil service." From that point on, if anyone was at the safe house during business hours, I would interpret it to mean they had nothing to do. They would then become my storm troopers for the day. We had a bridge in a rather dangerous part of the Province, at which bandits had been setting up a checkpoint and shaking down motorists. It had routinely been reported through the Americans that it was a Taliban checkpoint, but it was actually just a robbing crew trying to make some money at the Taliban's expense and reputation. Any *Tea Drinkers* caught loitering at the safe house would quickly find themselves lying in ambush with Sammy and me at the bridge, waiting for said robbing crew to show up. No one seemed to like that plan at all.

The next issue was that of paying informants. During the past two weeks, I had been asked repeatedly about how we were going to pay them for their reporting. My new policy was that we would not pay for information—we only paid for results. If their snitches wanted to spew out some lies and get paid thirty dollars, I would hook them up with the young military kids doing intelligence work on the prison camp. They didn't care about whether or not it was good information, because they just needed to write their reports. But, anyone who worked for us would only get paid after they produced tangible results. *Dope on the table.* If the information didn't pan out, they received nothing.

I reiterated to them we were a proactive unit, not a reactive one. You could either be proactive, or you would find yourself directing traffic at a checkpoint in the hot sun. I was determined to produce an effective unit that made a difference. If you were merely a *Tea Drinker,* start looking for a new home.

While this was an Afghan unit with no real direct subordination to anyone but the Afghan Ministry of the Interior, it was being funded with U.S. tax dollars. The average citizen in the U.S. was having a ton of money extracted (extorted) from their paychecks, even though times were hard there. Their sacrifice was what paid all of our salaries. If it killed me, the average Joe in the States would get his money's worth out of this crew. We would all earn our money. I felt compelled to do a good job, and felt a responsibility to both the American and Afghan people. Screw the formalities of "what it means to be a mentor." We were all cops. I looked at them no different than I would my American counterparts. To me, that was showing real respect. To coddle them in accordance with corporate expectations, would have been perceived as me thinking they were beneath me or less than my equal. I treated them like brothers.

At the conclusion of the meeting there were no questions. Instead of fueling the fire at the time, I elected to depart the safe house and let them stew over it for a while. I didn't know exactly what the outcome would be, but I didn't care. I would rather run a team of four people who were motivated, had the right mind set, and had a sense of responsibility, than run a fourteen-man team of non-hackers.

As Sammy and I went to the chow hall back at the prison camp, meetings were being held at the safe house. While I wasn't there, I bet it was eerily similar to when my compatriots and I were standing in the parking garage, trying to see who had a quick case we could put together.

● ● ●

The next morning at around 8:00 a.m., as we enjoyed some free food at the chow hall, Sammy received a call from the colonel.

"He says to be ready for a mission. They don't know exactly when or if it will go. He'll explain more when we get to the safe house."

"Sounds like our Come to Jesus meeting may have done some good," I said, with a mouth full of food.

Sammy and I hurriedly choked down the chow hall grub and headed back to the Paladin compound to get ready. As part of our readiness ritual that would become routine, we filled the entire hatch of the Forerunner with bottled water. On the prison camps, there were pallets of bottled water lying around. Gone were the days of filling canteens. You were still issued canteens for some reason but there was no potable water source in which to fill them. You would have had to use the bottled water to fill a canteen, but that made no sense. So, why were they still being issued out? Canteen manufacturer X still needs to make that money from the government contract.

We filled the back of the Forerunner just to the point where we could still pull the hatch cover over our cargo. We didn't want the gringos manning the gate to question us about where we were going with all that water. There was a lot of theft going on from the prison camps, so they should indeed be concerned. They just didn't need to be concerned with our business.

When running missions with the Afghans, it behooved me to bring along enough water for everyone. While it was far from being my duty, they would all look to the lone gringo to provide food and water. As long as they had water and something to eat, they would work all day and all night. While I love them all as brothers, they do not forward plan enough to bring their own rations. They just assume they will be on the mission for a couple of hours and be back at their base in time for chow. *No problem.* Therefore, if the mission runs long, my unprepared brothers get hungry and unmotivated. While it was a fight in itself teaching them readiness, I made sure to "steal" enough bottled water from the gringos to accommodate everyone. I started to bring cases of MREs (meals, ready-to-eat), but I wouldn't wish those nasty things on my worst enemy.

It was easier for me to carry a small amount of Pakistani rupees. In Nangarhar Province, Pakistani rupees are preferred over Afghani. The money was courtesy of the Bank of Mark. All I had to do was reach into my pocket to make a withdrawal. I could always buy bread, beans, and rice from the local villagers if I had to feed the troops. It was a small amount of money out of my pocket on the food, a small amount of money out of the U.S. government's pocket on the water, but it amounted to a lot being accomplished during a successful mission. It was pennies on the dollar for what the U.S. was spending on absolutely non-productive gadgets, technology, and personnel, in programs which yielded no tangible results.

I grabbed my rifle and ditch bag, along with my personal cameras. The troops all had small digital cameras but I wanted to make sure we always had good shots of our work. I carried a Canon Digital EOS Rebel that took great pictures. I also had a small video camera and a compact digital camera as a backup.

We fired up the Forerunner and headed toward the fuel pump. After topping off the tank, we stopped off at our shipping container in order for me to transform from gringo to Afghan. We made it out of the gate without any problems and arrived at the safe house twenty minutes later.

It took five minutes to get through the standard greetings, hugs, and handshakes. We made it up to the colonel's office where he and the case agent were having a discussion. They began to explain to Sammy what was going on.

Agent Z (full names are withheld for the obvious security reasons) had been contacted by a confidential informant (CI) at around 8:00 a.m. that morning. The CI had provided time-sensitive information.

Agent Z was in his mid-forties with a scruffy appearance. While he turned out to be one of the top producers of the team, his explanations and mannerisms brought him across to others as being an idiot. He was far from being an idiot. He was just an old country bumpkin from way back in the sticks. Sammy explained that Agent Z was what the Afghans referred to as a *Greasebon*. They were the poor, hill people who were farmers and hard workers, but had very little formal education. Some of them lived so far up in the hills that they still thought the Russians were in Afghanistan. Some thought the king was still in power. They had little to no contact with the outside world so it didn't matter anyway. Agent Z was from among these hill people. I related to him well, because I'm also from the country. While not Harvard educated, Agent Z had the worker mentality and the sense of responsibility to earn his pay and serve the people. I'll take a hundred of him, before I take a lazy educated one with a college degree.

Agent Z had been working with this particular informant for approximately three weeks on a narcotics investigation. He had actually done very little legwork on the case because of the dynamics of how it had to unfold. According to his informant, it had become public knowledge in his village that three brothers were involved with trafficking loads of hash. The loads would go from Nangarhar Province to a Taliban commander in Pakistan, known as Mangal Bagh. The people of the village had grown tired of the three brothers, known as Majur, Abdul Rahman, and Mudir, and did not approve of their activities. However, the people were deathly afraid of the crew because they had killed several people who had spoken out against them. The three brothers and their crew did whatever they pleased, whenever they pleased, and acted as if they were above the village elders. This did not sit right with the people. Something had to be done to regain order.

The informant didn't know exactly what residence they were using to stash (store) the drugs, but he vowed to find out. The CI advised Agent Z to be ready once he called. The informant indicated he would not call unless he had eyes on the location, as there was no need to waste anyone's time. If the call came in, it would be a solid case.

There were five different routes into the village. The informant had enlisted the help of several villagers and had established twenty-four-hour surveillance on these routes. They sat up on the trails non-stop for three weeks, waiting on the mule train carrying the hash to come through. Agent Z advised that last night at approximately 8:00 p.m., a mule train had passed one of the villagers on surveillance. They were loaded down with rice sacks and the smell of hash saturated the air. The villager successfully followed the mules to a compound inside the village and got eyes on the door. The mules departed approximately an hour later, without the bags. It was ready to be taken down.

The informant couldn't call Agent Z at the time because there was no cell service in his village. He had to wait until morning to hike out to cell phone coverage. According to the informant, the load of hash would be moved as soon as night fell. "Fresh horses" would arrive to take the hash to the next stash house in the journey toward Pakistan. If we wanted to interdict the load, we had better get over there before nightfall. If the load began to move, the informant wouldn't follow it and had no idea where the next stash house was.

After discussing the information and the background of the informant, Agent Z and the colonel were convinced the information was good. The village was inhabited by a handful of Taliban, but had mostly good people living there. No Americans traveled to the village so the threat of roadside bombs or mines in the road would be relatively low. The informant was well known to Agent Z and had no motivation to provide false information. While the informant and his crew were obviously looking for a bit of financial reward, their main motivation was to handle internal village business, without getting a bunch of innocent people killed.

The operation was a go. We needed to get to the PHQ in order to meet with General Ayoob Salangi, who was the police chief of Nangarhar Province. It would take some time after he issued the order, for his people to get geared up and assembled. That was the crutch.

I told the colonel to make sure the team leaders conducted equipment checks while we met with the general. Once we returned from the meeting, the two of us would conduct an inspection. On the way out of the building, the colonel ordered the team leaders to make sure everyone was ready to roll once we returned with the details.

At around 11:00 a.m., we were all sitting in the general's office drinking tea. On this first operational meeting with the general, I explained what we had and what we needed. This would be the beginning of a good relationship between the general and me. While the Americans typically conducted raids within his province and then told him about it later, here was an American coming to him for help. Here was an American who had just briefed him on all the facts and circumstances without keeping secrets, and showed him the proper courtesy and respect for his position. After I broke it down for him, he asked me how many people we needed.

"Sir, that's your call. If you tell us we need twenty people to take down a house in that village, then twenty is what we need. If you say thirty, then give us thirty. It's your backyard. We're just playing in it. You know the area better than any of us."

The general smiled and pushed a remote control doorbell button on his desk. All of the commanders I had encountered in Iraq, and the commanders I would encounter in Afghanistan, had the old remote control doorbell. After ringing, a young lieutenant was in the room ready to take direction.

The general barked some orders to the lieutenant who immediately disappeared. Per Sammy, the general told him to contact the colonel in charge of the Counter-Narcotics Division, and have him assemble fifty men for a raid. They had two hours to be ready to roll. In the essence of time, we thanked the general for his assistance and excused ourselves. We would get together with him soon, to discuss the results.

The Counter-Narcotics Division was located in a small compound, within the overall compound that made up the PHQ. We stopped off to speak with the counter-narcotics colonel to give him a quick brief on what was going on. He assured us he and his men would be ready to roll as directed. We agreed to assemble all personnel involved, at his compound, no later than 1:00 p.m.

Back at the safe house, we found the troops drinking tea on the front porch. Somehow, I knew that they were less than ready to conduct a drug raid.

"Have everyone stand beside their respective vehicles and prepare for an inspection," I told Sammy.

Between Sammy and the colonel barking orders, the men were finally assembled by the Forerunners. Many were still clinging to their tea glasses.

The first thing I checked was a vehicle itself. A quick turn of the key revealed the fuel hand showed half a tank.

"What the hell is this?" I yelled, in no particular direction. "I go through too much trouble to shuttle fuel cans out here, so these vehicles are ready to roll at a moment's notice. Get it filled up!"

I was pissed. Almost every time I rolled out of the prison camp, I was armed with three or four full fuel cans. I went through the trouble of covering them up much like the water, in order to not be accused of stealing when transiting the entry control point. The lids on the cans leaked like crazy and fuel was constantly sloshing into the back hatch of my ride. By the time I got to the safe house, I was high as a kite from sniffing gas fumes. One stray spark and I would have been a barbecued gringo. The fuel cans were emptied into a fifty-five gallon drum so we had a fuel reserve. It was the only way to ensure our vehicles remained in a constant state of readiness. The prison camp was frequently locked down and closed to vehicular traffic, due to "security threats." Because of this, we couldn't assume we could always get fuel at a moment's notice. I gladly went through the pain of the process, but they had disrespected my efforts by not being prepared.

The young soldier scurried to get the fuel situation squared away. The vehicles were part of his responsibilities. I checked the hatch area for the emergency equipment and found none. My disappointment and anger intensified. A quick check of the four agents standing by the vehicle revealed that only one person was armed with his radio, and only one was armed with a rifle. A cursory inspection of the other two vehicles revealed the same situation. *What the heck kind of a crew am I running here? There's no one to blame but myself*, I thought.

I'd already harped about readiness but apparently no one paid attention. They looked like a bunch of Keystone Cops. They weren't even ready to raid a donut shop, much less a stash house with a million dollars-worth of Taliban dope and armed bad guys. There was no way anyone was rolling out like that, especially since we would be in the company of other police officers, junior to us.

"Damn it! What the hell have they been doing for the past hour? OK, let me treat everyone like children."

A few of the agents spoke enough English to understand what I was saying. The ones that didn't, still understood the message. They knew exactly what the problem was. It was a typical case of Afghan *no problem* mentality. *No problem,* let's roll out into a potential firefight without being prepared.

"Get Agent F to open up the equipment room. We're wasting time."

As soon as Agent F opened up the equipment room, I started throwing much needed equipment out into the hallway for agents to load into the vehicles. First aid kits, aircraft marker panels, evidence collection kits, flexible handcuffs, extra batteries…

I looked at the radio charger and noticed it was chock full of radios. There were more radios sitting next to the charger. Now, what good could a portable radio do on a raid, if it was back at the safe house?

"Sammy, you tell the colonel that I want every one of these radios issued out. I don't care if there are more radios than men going on the raid. Issue them all out, along with all of the spare batteries."

The colonel looked at Sammy with a puzzled look as Sammy relayed my "request." A conversation ensued, which resulted in the colonel reluctantly telling Agent F to issue out the radios.

I looked at the shelf below the radios and observed the stock of GPS units that had been issued to the team. There had been fifteen GPS units issued. We had gone through a lot of trouble to obtain one for each agent. The other teams in the program only had five. I was looking at fifteen GPS units collecting dust on a shelf. At the time, I erroneously assumed they all knew how to operate the new GPS units.

"Have every man that's going on the raid to come and draw a GPS unit. That's not negotiable."

As I continued through the equipment room, chunking items out into the hallway to be loaded, I stumbled onto the weapons rack. It was full of rifles. I remembered that only a few agents had theirs with them.

"Have every man draw their rifle and six mags."

"Sir, I'll tell them, but I'm going to tell you what they'll say. They'll say that the rifles are no good and will jam after the first round. They don't like the rifles and don't believe in them. They are pieces of junk, sir," Sammy said.

"I know what they think, but right now they're the best we can do. If they jam after the first round, that just means they'll have to make that one round count. We'll kill at least fifteen from a distance, and then they can use them as clubs. But they will carry them."

I realized that Sammy was just trying to explain the dynamics and the reasoning behind the men not wanting to carry the rifles, but we didn't have time for a debate. Just tell them to come draw one. As I've described before, the rifles were cheap knock-off versions of the coveted AK-47. I wasn't arguing that they weren't shitty, but they were all we had.

I realized poor Agent F was trying to keep up with the inventory that was rapidly leaving his precious, organized equipment room. The colonel also had a look of worry about whether or not Agent F was accurately accounting for the gear.

After a thirty-minute flurry of action and preparation, I went back through the vehicles. They were full of fuel and had all of the necessary safety equipment. The men all had rifles, handheld radios, GPS units, and additional magazines of ammo. I will concede that on this raid, the magazines for the rifles were only loaded with less than ten rounds each.

Now, we looked like an elite, specialized unit.

The colonel called everyone into the conference room for a pre-raid brief. After he and Agent Z went over the operational plan, we still had a few minutes before we needed to depart for the PHQ.

Through Sammy, I explained to the men that I wasn't being a tyrant, just to be a tyrant. They had been given that equipment for their safety. They needed to make sure every time they rolled out on a mission, they were prepared. Never go out half-cocked because you'll get caught with your pants down. I assured them that before any mission, I would be conducting the same inspection. They might as well get used to the routine. Part of the reason for the lack of preparedness was that complacency had set in. The men were older and more experienced than troops fresh out of the academy. It took a lot to get them excited. But as the saying goes, complacency kills.

We arrived at the PHQ to find several Ford Rangers (standard vehicle for the Afghan National Police) lined up and ready to go. Every vehicle was armed with a PKM machine gun on a mount in the back. PKM is Russian for, Pulemyot Kalashnikova Modernizirovanniy, which translates to, Kalashnikov Machine Gun Modernized. It was designed by Mikhail

Kalashnikov, the same man who designed the AK-47. Rocket-propelled grenade (RPG) launchers and additional RPG rounds littered the back of the trucks, along with additional cans of ammo for the PKMs. The convoy commander and the counter-narcotics colonel were conducting a briefing with around thirty-five uniformed officers. Another ten officers in plain clothes were also in attendance. The uniformed officers were standing in formation and almost appeared to be at parade rest. Everyone involved was focused on the person speaking.

I was impressed. This was counter to everything that had been, and was being reported about Afghan Security Forces, as a whole. These guys were nothing but professional, so far.

I noticed that a woman in a blue *burqa* was standing off by herself behind the group. Sammy explained that she was a counter-narcotics officer and would accompany us on the raid. If there were any women or children to deal with, that was her job. She had a police uniform, but this wasn't Kabul. In Nangarhar, she rolled out wearing the *burqa*. Nangarhar Province is the equivalent to the Bible Belt in the United States.

After the operational briefing concluded, the stage was set. It was time to roll, before the dope rode off into the sunset on fresh horses. At 2:00 p.m., the convoy departed friendly lines of the PHQ. Traffic cops blocked the roadway and the mass of traffic, to allow the convoy to enter. A plume of dust was kicked up as our Forerunners led the way, followed by ten ANP Ford Rangers with blue lights flashing. Officers manned their machine guns as if battle was imminent. The small task force we had assembled probably could have successfully invaded Mexico.

It also reminded me of the start of the Talladega 500. I could swear that a couple of ANP trucks passed one another in a jockey for track position, even though we were all going to the same place at the same time. While the mission was indeed time sensitive, the drive to the target area ended up being the most dangerous and thrilling aspect.

It was a trek to get into the vicinity of the target village. The journey was delayed by the fact that we stopped to pick up additional ANP officers from the local district center. We never called and told the local district chief where we were going, ahead of time, for operational security purposes. The general may have told him, but regardless, we never got burned on a deal because of the local cops. The district chief was always willing to help and would send a truck full of men who knew the area.

Once we left the paved road, that's where things got real. Fast. It wasn't a problem to set up a bomb on the side of a paved road, but it was too easy to dig a bomb or a mine into a dirt road. We were riding in vehicles that would be blown into a thousand pieces by even the smallest anti-tank mine. There was no way to defend against these threats other than by using commonsense and hopefully having a bit of luck on your side. It didn't even matter what number vehicle you were riding in. Sometimes, the last vehicle in a convoy would hit the mine or the bomb.

Psychologically, to overcome any fear, Sammy and I developed a ritual. Once we left the paved road and entered no man's land, I would say, *Everybody wants to go to heaven,* and Sammy would respond by saying, *But nobody wants to go now!* In unison, while slapping a high-five, we would shout, *Fuck it, let's go!* Laughter would always follow and lighten the mood. While that might be one for the shrinks to evaluate, it was our way of saying *screw you,* to death and the Grim Reaper. After that, Sammy and I didn't worry ourselves with the inconvenience and thought of possibly being blown up by a mine. That was left to chance anyway. So, why worry about it? We would always lead the way into missions, so no one could ever cry about the danger of getting blown up on the way in. As a side effect, no one ever seemed to want to ride with us.

We pulled off at a secluded location overlooking the village. In total, we had amassed a force of over fifty men. The plan was for our undercover agents to transport the informant into the village and identify the target house. The informant would be concealed underneath a blanket, in the hatch area of one of our Forerunners. The convoy commander would watch the Forerunner through binoculars from a distance, out of sight. Once the informant identified the correct house, the agents would covertly signal to the convoy commander. After they (undercover agents and the informant) departed the area, the convoy would converge on the target house and establish a perimeter. Not a bad plan, and they were the ones who came up with it.

We watched the Toyota Forerunner travel down the dirt road from our vantage point. It disappeared from sight for a few minutes and re-emerged as it entered the village. It bounced along, attracting no attention from the villagers.

In the Forerunner, Agent Z told the informant to get ready. The informant rose up from under his blanket, just enough so that he could see out the side window of the hatchback. If anyone could see through the tinted windows, they would only be able to see a set of eyes peering back at them. A few minutes later, the informant got excited.

"That's it!" the informant said, as he pointed.

No one in the car could see him point, but Agent Z and the other two agents understood which house he was referring to. They had the place positively identified.

Agent Z pulled the Forerunner down the dirt street about fifty meters and parked on the side of the road. He exited the vehicle and began walking back toward the target residence.

From our vantage point, the convoy commander watched intensely. It was apparent that the target house was along that particular street, but he still wasn't sure which one it was. It soon became clear.

Agent Z walked in front of the door to the target residence, began to cough a bit, and in doing so, "dropped" his *pakol*. That was the signal. Agent Z picked his hat up from the dirt, turned, and calmly began to walk back to the Forerunner. I had to laugh. Cops all over the world are the same. In America, if you're about to sell drugs to a guy who's wearing a hat, you're probably about to get screwed. If he takes off the hat after you hand him the drugs, he's a cop. Put your hands behind your back and lie on the ground. The take-down team will be there shortly to arrest you. The old "hat off" as a take-down signal has been used for decades.

The convoy commander gave the hand signal to get ready to move. By the time he jumped into the lead vehicle, everyone was waiting solely on him. As the convoy began to roll as one large unit, Agent Z and crew turned off of the target street and disappeared into the village. They had done the hard work of finding the place. Now it was time for the muscle to do their job.

As the convoy jumped, bounced, and dodged holes in the roadway, it had obviously caught the attention of the villagers. The element of surprise was up at that precise moment. The several minutes it took to reach the target street didn't matter in the grand scheme of things. A bad guy couldn't exactly put a ton of hash in his back pocket and run out the back door. The only thing that could happen is either the bad guys would run, or worse, stand and fight. If they ran, we would identify them through the investigation and find them later. The main objective was to knock off the load and keep the proceeds out of Taliban hands.

Even before the wheels on the trucks stopped turning, the convoy commander was yelling commands over the radio. Asses and elbows jumped from the back of the pickup trucks and started running. Without error, our small army swarmed the exterior of the location. Within a matter of seconds, an impenetrable perimeter had been established around the qalat. Uniformed officers were positioned high and low. Machine gun positions were established at two locations, in addition to the machine guns mounted in the back of the vehicles parked on the street. Several uniformed officers were armed with RPGs, just in case. We had the dope contained, and hopefully, we had some rats in the box as well.

The Afghan National Police conduct business much differently than the American military. Of course, the objective of law enforcement is not about killing people. While we could have rolled in strong on the target house, it wasn't necessary. It wasn't as glamorous as kicking in the front door or blowing it off its hinges with explosives. There were no flash bangs deployed to deafen any young children.

After the perimeter was established with overwhelming firepower, it was time to negotiate with whoever was inside. Now, this approach doesn't work well with drug cases in the States. The reason is that if you give the bad guy time, he'll flush as much dope down the toilet as he possibly can. If you give him enough time there won't be any dope left in the house, unless it's a big stash. The second possibility is that you give the bad guy time to prepare for your entry. He has time to barricade himself or get ready to shoot. Therefore, the combination of speed, surprise, and aggression is the concept. You try to catch him off guard before he can get to his guns, by using the element of surprise.

In Afghanistan, when the entire village knows you're coming for over five minutes, there is no element of surprise. You assume the path to the target is mined and that there are bombs alongside the roadway. You assume the bad guy's got explosives ready to do you harm. You already know there are AK-47s within arm's reach. RPGs are as plentiful as bullets so you assume you're facing those as well. Those factors are a given. Why give the bad guy a heads up for over five minutes that you're coming, and then do what he expects? That's what the U.S. military does. They have no concept of stealth. Maybe they think they do, but it's impractical. Whether you arrive via helicopter or hump in from somewhere, the villagers know you're coming. You can hear a helicopter a mile out. But, they will still hit a place strong because that's their method. No wonder they end up killing three women, two dogs, and a donkey on every raid, while their bad guy is nowhere to be found.

After the box (perimeter) was established around our target, the old prosecutor got out of the last vehicle and walked up to meet with the convoy commander. This was another main difference between us and the U.S. military. Afghan cops follow Afghan law. They operate within criminal procedure, just like American cops do. I'm still not sure which set of laws the U.S. military operated under. I'm still not sure what gave them the authority to raid a citizen's home. It's a sovereign nation, I think. Under Afghan law, you have to have a judge's approval to kick in a door. I don't think an Afghan judge would willingly issue an order, which allowed the gringos to blow someone's door off of the hinges, in the middle of the night, with women and children inside. But that's another topic for discussion.

Technically, the cops can, with the prosecutor's approval, hit a place strong and then obtain a search warrant after the fact. As long as there are exigent circumstances (like the dope was about to leave the location) this is acceptable, just like in the States. But, why hit a place strong and have it potentially get messy when you can be civil about things?

The old prosecutor conferred with the convoy commander and our colonel. The convoy commander then yelled over the wall for the occupants to open the door and come out. He told them we were the police and that we had the place surrounded. There was no need to be ugly about things. The door opened, and a man known as Major, emerged. The old prosecutor, the convoy commander, and our colonel had a conference with Major. Sammy would later explain that they were asking for his consent to search the residence. Similar, but different in tactics, to doing what we call a "knock and talk" back in the States. In the States, it has to appear very non-confrontational when you ask for consent. No guns drawn and no overwhelming show of force. It has to appear as if it was absolutely voluntary or you'll have your evidence thrown out at a suppression hearing. As there was a small army surrounding the place, the asking for consent was a bit different than in the States. I had to remind myself that this was Afghanistan. I wasn't in Kansas anymore. Even if Major had not consented, the prosecutor was already on board with the investigation. We weren't leaving without the dope.

A few minutes later, the four actors walked into the compound. The counter-narcotics officers along with several uniform troops followed. After clearing the compound, only one elderly woman was found to be inside. Major agreed to the search and the prosecutor gave his approval to start looking. There were no other women in the compound and therefore no cultural issues to deal with. The female counter-narcotics officer sat with the elderly woman while the scene played out.

In one of the small rooms, we hit the jackpot—thirty-eight rice sacks packed to the gills with high quality hash. The brownish substance in the sacks was as hard as rock. It was like picking up bags of concrete.

As nightfall was near, there was no time to screw around. There was no doubting the reason we were there. We were stealing from the Taliban. While the majority of the villagers were quietly behind us, it would only take one person to get to cell service and alert the local Taliban commander. They would have plenty of time to set up on us as we departed with their precious cargo.

It was a good thing we had brought so many vehicles. The bags of hash weighted down the Ford Rangers and barely left room for the men to sit on top of the sacks. As darkness set, we departed the village with one bad guy and thirty-eight sacks of hash weighing in at 3,395 pounds.

As the string of headlights traveled in unison, driver's strained to see through the dust generated by the preceding vehicles.

"Hell yeah, this went down like clockwork." I said. "This is a good lick for us."

"Excellent work," said Sammy.

"If we can get the hell out of here without getting hit, it will be a good day," I added.

A couple of hours later, we pulled into the PHQ with our haul. I felt a sense of accomplishment from everyone involved. Conducting proactive investigations was just something they weren't used to doing. They seemed to have the capability and the know-how, just not the proper motivation. All it took was Sammy and me forcing them to go to work.

I made it to the prison camp after midnight and began the hassle of getting through the gates. While the day shift guards had gotten to know me, the guys working nights seemed to rotate around a lot. I pulled up to the gate, turned off the headlights, and turned on the dome light while holding up my gringo ID card. I was promptly met by a young Afghan whom I greeted in Pashtu.

As usual, he inspected my ID card. While I came to realize that most of the guys guarding the camp couldn't read, he at least knew I was a gringo. The process then was to radio the American soldiers who were about a hundred meters back, manning the secondary checkpoint. But I knew this drill. The Afghans would radio for the Americans to come up and verify my identity, but no one would ever respond on the radio. I would sit outside the gate for twenty minutes, until they finally realized any attempt to contact the soldiers was futile. Maybe the radios didn't work, but at that hour, the soldiers were probably sleeping. I shouldn't say that because I really don't know. All I know is that at night, it took me forever to get back through the prison gates.

Twenty minutes later, they let me through to the American guard post. After a brief encounter, I was finally on my way around the airfield to the compound. As exhausted as I was, I dragged myself into the TOC in order to send out a quick e-mail. I had come to realize that if I didn't send out at least a brief e-mail taking credit for our work, someone else would take the credit for us. It would usually be an entity that had absolutely nothing to do with the case.

I typed out a brief e-mail indicating that my unit had conducted a narcotics investigation, which resulted in the seizure of approximately 3,395 pounds of hash and the arrest of one bad guy. The bad guy's name was "MAJOR (son of) ADAM KHAN (son of) DAR GHUL." A press release would be conducted the next day at Nangarhar Police Headquarters. A storyboard (PowerPoint presentation) and report would follow. I hit send, thus blasting it out to an extensive distribution list and then retired to my cubicle.

Little did I know that the guy's name would create a shitstorm of false assessments and ill-informed conclusions by a plethora of military folks. The next morning, a flurry of e-mails that I was cc'd on, contained discussions about the identity of the *Major* (rank) we had arrested with the hash. What agency was he with? Was he with the Afghan National Police? Or, was it the Afghan National Army? Or, maybe he was with the National Directorate of Security? The authors involved were "not surprised" at the arrest of a high-ranking Afghan officer. They were all corrupt anyway. The discussion went on and on. Nothing was discussed about the fact that ANP officers and my unit had just kicked the Taliban in the nuts and stole their dope.

I immediately sent out an e-mail reiterating the fact that Major was the guy's first name and not his rank. He was in no way affiliated with the good guys. As a matter of fact, I personally changed the guy's name to *Majur* on all the reports. I spelled it with a "U" to avoid any further misunderstanding among the gringos. Looking back, the situation was actually kind of comical.

Sammy arrived at the office and informed me that the public information officer at police headquarters had scheduled a press conference for around noon.

This was good. It accomplished a lot of positives. First of all, it established a precedent for how we would do business with the general and his officers. We would do the dirty work, up until the point to where we needed muscle. They would assist, and then take all the credit and enjoy the publicity from the operation. We were a covert unit, so we very well couldn't stand in front of video cameras and be interviewed. But, the information needed to go public.

The press release would show the people that the government was in charge. It would demonstrate that the police were capable of conducting proactive investigations and enforcing the law. It would also generate interest among the military folks who had yet to learn our name. While I was eager to get started on my report and storyboard, I wasn't going to miss the press event. We would stand quietly in the background of the crowd to ensure that our pictures weren't taken. Afterward, we would get a trophy shot with all of our agents who were involved.

When Sammy and I pulled into police headquarters, there was already a flurry of people inside the counter-narcotics compound. Reporters milled around while cameramen assembled tripods. Several high-ranking ANP officials were mingling among the visitors.

Sammy and I quietly stood at the fringes of the group, observing. I wanted to take some pictures of the event to include at the end of my storyboard. Several of our agents arrived and joined us.

A group of Americans walked in through the gates. There were at least four DynCorp police advisors and several U.S. soldiers. They had come to see what the excitement was all about. As it would become my habit, I didn't interact with them. I stayed in stealth mode to avoid any unnecessary questions or scrutiny. They walked past, as if Sammy and I were the chai boys.

After the arrival of the prosecutor, the seal of the shipping container (used as the evidence locker) was broken. The cameramen were allowed to take pictures of the thirty-eight bags sitting inside. The smell of hash filled the air. A set of scales was assembled in front of the evidence locker, while a counter-narcotics officer was armed with a ledger. Additional counter-narcotics officers had small paper bags at the ready, in order to take representative samples. Young ANP officers wrestled the huge bags out of the shipping container and onto the scale. One by one, they were weighed, at which point a sample was taken from each bag. The sample was placed in a smaller paper bag, for later analysis.

The bags were stacked on top of one another, where they formed about a four-foot-high wall. With the wall of hash as a backdrop, the public information officer began the interview by announcing the seizure and the arrest. He was then interviewed by several reporters, and ended up with about thirty minutes of footage to his credit.

The suspect, Majur, was then led out in chains and displayed in front of the wall of hash. This practice is customary in Afghanistan, and is standard operating procedure in several other countries, as well. In the U.S., we don't allow the media to photograph the bad guy together with the evidence. We, as Americans, think it's suggestive and could prejudice a jury. As Majur was positioned in front of the cause of his newly found incarceration, he began to wail like a baby. The counter-narcotics officers immediately shielded him from the limelight and escorted him into the building. The Afghans are very compassionate people. It wasn't personal to them, and they didn't look down upon Majur because of the incident.

After the press got what they needed and were beginning to pack up their cameras, the Americans found an opportunity to make asses out of themselves. As they discussed and speculated on how the Afghans had seized the load, a young staff sergeant decided that it would be a good photo opportunity and jumped up on the wall of hash. He kicked back as if he was watching television from his couch.

"Hey, take my picture!" the young idiot yelled.

As he was laid back on top of our evidence that could potentially sentence a man to a lengthy prison term, his buddies began taking pictures while laughing.

None of the Afghans thought this was funny. I found it highly disrespectful for these uninvited guests to be screwing around like children. However, a couple of cameramen from the media figured they would get a few shots of the village idiot, just in case. I was embarrassed as a gringo. I was embarrassed for them. The four DynCorp advisors also took the opportunity to get a trophy shot with our dope without asking.

OK, you've had your fun, now get the hell out of here. They finally left. I'm sure they had checked their watches and realized the chow hall was about to close. From the size of their bellies, they certainly were not going to miss a meal.

After the masses had cleared, the true agents who generated the case were able to get a few trophy pictures with the haul. We made sure that we were good on paperwork with the counter-narcotics guys, and then departed. Back at the safe house, I congratulated the men on their efforts. I didn't have a lot of time to hang around because everyone and their brother would be expecting my reports. Before I left, I decided to show them how appreciative I was, by announcing a pending get-together.

"We're going to have a barbecue to celebrate, as soon as we get a free day," I promised the team.

As Sammy and I rolled out of the gate, I thought of Mel and his methods of motivation. They certainly worked. *I'm going to need a bigger table,* I thought.

By the time I had polished up the report and storyboard, I was already reading more assessments about our seizure. Apparently, the village idiots taking pictures on our evidence had already published some inaccurate information that was making the rounds as well. It was par for the course.

I sent out the real story, giving our team the appropriate credit for our work. The response was a few sniveling remarks from within the military channels about not being notified of the operation. After a pretty good gig, all they could focus on was the fact that not everyone within the type-a, steroid-munching, weight lifting, chain of command, had been notified. They were offended by the fact that the Afghans (their country), specifically the Afghan police (their country and their jurisdiction), had conducted an operation without the express written consent (don't need permission; once again, it's their country) of the Americans who never leave their prison camps.

This mentality will paint the picture as to why the U.S. has been in Afghanistan for ten years and has accomplished nothing. Battlespace renters arrive with a mentality that they "own the ground" in which they've been designated on a map. They believe all decisions to be made are theirs. They live under the misconception that they control that area for their twelve-month stay.

As was typical with most of our operations to come, the Americans were only concerned that somehow, in some way, a policy they had in place was not followed. Never mind the outcome. Americans are only concerned that the method adheres to the U.S. way of thinking, because after all, we're always right.

Now, I was on the hook to figure out how we were going to pay our informant for the case. The Afghan process of obtaining operational funds from Kabul was broken. Even though CSTC-A had pumped $300,000 U.S. dollars into the Afghan Ministry of the Interior (MOI) bank accounts specifically for this purpose, we had yet to see a dime. The money would sit in the account until the minister and his cronies could figure out how to steal the money, without being absolutely blatant about it. The thing that pissed me off, was that no one from the American side would hold anyone's feet to the fire, and make them do what they were supposed to do.

"Once we give them the money, we can only advise them on how to use it," was the same old tired story from the U.S. military. Did I mention that people back in the U.S. were losing their homes while their tax dollars were being wasted by impotent policy and procedure?

If the minister was just going to sit on the money and eventually divert it to his bank accounts, then don't give him any more money. Regardless of the problems with upper government officials in Kabul, the bottom line was, the people with boots on the ground didn't have funds to pay their informants.

I approached one of the brigade intelligence officers, a captain, about getting funds to pay our informant from the small rewards program. SRP funds were typically utilized to pay for weapons and explosives caches that had been seized. We used it on all of our caches without a lot of hassle, but this was a dope case. The captain was a non-social individual, about thirty years of age. Overweight, with glasses and a bald head, his appearance emulated his work ethic. In layman's terms, I found this gent to be as lazy as the day is long. As I walked into his cramped office, I immediately felt as if I was bothering him.

"Sir, did you get the storyboard and report I sent out on a deal we did?" I asked, quietly.

"I probably got it, but I just delete all those reports you cops send out," he advised.

That didn't surprise me at all, and I really didn't care about his opinion anyway. I needed the fat *fucker* to do one thing for me, and that was to approve some funds for our snitch.

"I was wondering if you could help us out with paying our informant on the case. Since it was a dope case, I'm not sure if our normal channels through SRP will pay."

"Who was the source working for?"

"He was working for my team," I said. I knew where this line of questioning was going.

"If he wasn't working for HCT (Human Intelligence Collection Team) then we probably can't pay him."

First of all, this guy was just plain lazy and didn't care. Maybe he had been deployed five times in five years. If that was the case, then I understand his lack of motivation. At the present time, it was apparent that he wasn't going to help out this nasty old contractor standing before him. I elected to excuse myself without creating any more attention to the situation.

"Well, if you can think of a way to help us out, please let me know. You take care, Captain," I said.

I slipped out of his office while cussing him under my breath. This gentleman was the type of person who only did things when either ordered, or to stay out of trouble.

I went to the major over Task Force Paladin to try and get him to help. He was a great guy but was always being pulled in too many directions. I hated to burden him with a menial request, but I had nowhere else to turn. There was a good snitch on the line, who deserved some cash rather than excuses.

After meeting with the major, he advised he would make an inquiry to see if he could get someone to roger up the funds. That was good enough for me. He was a Marine. If a Marine says he is going to handle something, you can take that to the bank (unlike his army counterparts, who cared only about the date they were leaving Afghanistan).

A couple of days later, the major reported that he had been advised of what was needed to get our guy some money. The brass, on up the chain, advised that as long as the case was connected to insurgent activity, they would cough up some funds. If it were strictly a dope case, they wouldn't pay. It had to be connected to anti-Afghan forces. I would need to show some documentation in reference to this.

I thanked the major for his efforts. In all of the confusion and chaos of conducting the operation, we hadn't had time to sit down with the informant and get all of the details of the organization. We only had enough time to raid the stash house and steal the dope. The press release tied up the next day. We needed to call in the snitch for a detailed interview, as soon as possible.

At our request, the informant made his way into J-Bad. Our agents scooped him up and shuttled him over to the prison camp. Sammy and I met them at the gate in order to escort them in. While I had trained our agents to search anyone before we let them in our vehicles, I made them put the snitch out on foot, in order to walk through the gate. As he walked through, he was searched again by the security guards. With me standing there watching, they did an extra thorough job every time. This gave me total peace of mind that we weren't bringing a guy onto the prison camp who had a grenade taped to his nuts.

I had endured the trouble of obtaining badges for all of my agents, in order for them to enter and exit the prison camp on their own. They were police officers who were carefully selected and vetted, and were party to high-risk operations that usually benefitted the Americans. They were party to more sensitive information than anyone on the prison camp. You can trust them. However, they were treated with the utmost disrespect every time they attempted to come onto the prison camp, without an escort. Even with the required badges, they were routinely detained because they were found to be carrying weapons. An Afghan in Afghan clothes, with a weapon, must be the Taliban. The gringos seem to think they all have bombs in their pockets as well.

I would explain to no avail, that they were undercover police officers. They had the appropriate badges for the camp. They had police identification cards. They even had a gold badge just like an American cop, but it wasn't good enough. Plain clothes and Toyota Forerunners must equal insurgents.

The illogical thing was that if anyone pulled up to the gate in a Ford Ranger pickup truck, who even resembled ANP or ANA, they automatically let them in. Never mind the guy in the back, manning the machine gun. Since he had a uniform on, the American soldiers didn't bother checking identification cards. Just come on through with that PKM and those RPGs. My agents had all of the necessary documentation, and then some, but would get screwed with every time. It was easier for me to drag my ass up to the gate to ensure there were no disputes.

Back at the Paladin compound, we tried to make the informant feel at home. I rounded up some Cokes and a can of mixed nuts from the break room for the crew. The informant was dressed in traditional garb but had an air of distinctiveness about him. After studying him for a few minutes while he was speaking, I came to the conclusion he was no farmer. His mannerisms and speech indicated education and intelligence. It raised a level of suspicion that I needed to explore.

As the agents spoke to the informant, I asked Sammy to inquire about what the gentleman did for a living, and to find out a bit more about his background before we continued. I needed his biographical information for my report, anyway.

What I truly wanted to rule out was whether or not this guy was Pakistani ISI, the equivalent to our CIA. While he had been searched thoroughly by our agents and by the guards at the gate, it's too easy to conceal a small voice recorder. We were the only gangsters in the room and would not be played.

Sammy, Agent Z, and the informant had a brief conversation. Sammy translated that the informant had known Agent Z for several years. They knew one another's family and he was definitely not ISI. What he had, was a master's degree in economics. He had been an economics professor at the university in Kabul during Russian times. He hadn't taught since the Taliban took control of the country. Now, he was merely the smartest farmer in the village. He had been the mastermind of the surveillance operation, which resulted in detecting the load and finding the stash house.

The informant elaborated that Abdul Rahman (Majur's brother) was the main guy in the drug organization. Abdul Rahman had fought with the Mujahideen against the Russians and then later aligned himself with the Taliban. He was also involved in the fighting at Tora Bora when the Americans first came to Afghanistan. He was coordinating the loads of hash for a higher-level Taliban commander, known as Mangal Bagh, who was in Pakistan. Mangal Bagh would inevitably use some of the profits to finance operations in Afghanistan against the Americans.

The informant went on to elaborate that we all had been lucky on our approach into the village. Word later spread around the area, that some of Abdul Rahman's Taliban friends had observed the convoy traveling toward the village. The fighters had "remotes in their hands," ready to detonate roadside bombs against our convoy. They hesitated, because they mistakenly thought the convoy was a personal security detail for Hajji Zahir. They thought that Hajji Zahir, Abdul Rahman's good friend, was coming to the village to personally welcome him back from a trip to Pakistan.

...Hajji Zahir was a well-known smuggler who had been elected to parliament, which was a typical story in Afghanistan. It takes money to get elected, much like everywhere else in the world. Who had the money? It was the warlords, the smugglers, and those on the CIA's payroll.

We were lucky, according to the informant, because we took a different route out of the village. The handful of Taliban fighters were not prepared to detonate IEDs on that side. They were scared to attack the convoy with small arms, because we far outnumbered them and were carrying more firepower. Overall, the people were happy about the operation and were surprised that the police were not afraid of Abdul Rahman's status or reputation. The informant continued with details of intelligence value which I would include in my report. After the interview came to a close, I decided to address the problem of the money, and came clean with the informant.

"Right now, I'm working on securing funds to pay you and your crew for your efforts. I'm dealing with Americans and their paperwork. I can't tell you when or even if I'll be successful, but I'm trying. Meanwhile, we've got some food, clothes, and blankets that I want you to take back with you. It's the best I can do right now."

The informant didn't seem to be concerned about the money at all. He was very thankful that he would be taking back basic necessities. In reality, it wasn't about the money. Those crooks had gotten out of control and had upset the balance of the village. They had to be set straight, and that's exactly what happened. The money would be nice but it wasn't the underlying issue.

As we headed toward the gate of the prison camp, the informant advised he would be working on another deal as well. No need to waste our time with the details. If he called, we should be ready at a moment's notice. After this case we would always jump when the guy had information. He was the real deal.

CHAPTER 4

Gaining Momentum

The hash case gave the team confidence. It gave them credibility among their peers within our program and also from within Provincial Police Headquarters. The local cops at the PHQ knew only that the team was a secret unit from Kabul, being run by a "crazy American." While we may not have brought down the underworld by knocking off a little bit of hash, we put a small dent in it, psychologically. It gave the government credibility within the community, and created a stir among some of the bad guys we were keeping tabs on.

While the dust was still settling, Agent S received a call from one of his informants. It was important he meet as soon as possible. He had information on the whereabouts of a kidnap victim, who was being held hostage for ransom. The informant advised it would take him until the next day to get to Jalalabad, and for us to be ready to receive him.

The informant called from the bazaar at around 11:00 a.m. Agent S, the colonel, Sammy, and I rolled out to meet him. Agent S made contact with the guy and ensured that he wasn't wearing a suicide vest, with which to blow up the American. This was easily accomplished in the Afghan culture, because everyone always hugs as part of greeting one another. The agents would merely make a few extra inquiries disguised as hugs to ensure the person wasn't armed. No one thought any different of it, and no one was ever offended by us questioning their loyalty or motives. Elementary level police work teaches you to search the people you deal with, or else you might find yourself dead.

We all herded ourselves into one of the local restaurants which was crowded with patrons. We elected to secure a private room, upstairs in the hotel. When you think of the word hotel, you obviously think in a western sense. This hotel merely had bare rooms with mats in them. The bathroom was just a Turkish toilet in a small closet located in the hallway. It was community property for all to share. No amenities here, but a hotel nonetheless. It would serve its purpose for the meeting and also give the snitch a place to stay, should we decide to keep him for the night.

After being seated, the waiter arrived with water in a metal pitcher and an empty bowl. This was so you could wash your hands. He went around the room until everyone had a chance to wash up.

The colonel placed an order for our meal and the waiter hurried off. The door was shut to ensure there were no eavesdroppers and the meeting began. As Agent S and the informant conversed, Sammy kept me in the loop.

According to the informant, around two months prior, there had been a dispute between two rival groups of drug traffickers in Achin District. The dispute had somehow involved 161 kilograms of opium that had changed hands. Somehow, in some way, someone was still owed 2,500,000 Pakistani rupees (around $30,000 U.S. dollars). In order to secure the perceived debt, the offended group decided to kidnap a family member of the opposing team.

The kidnap team sprung their plan and abducted a young man driving his donkey cart near a village. The poor guy was only about twenty years old. To make matters worse for the victim, it was determined that he wasn't related to anyone involved. He was just passing through the area on his way to the bazaar. It was a case of being in the wrong place at the wrong time with very bad luck.

The bad guys had taken the kid, unannounced, to Narghosa Village, to a residence of distant acquaintances. They instructed the three occupants of the house to keep the kid under wraps and to make sure he didn't get away. While the folks who lived in the house didn't want anything to do with the deal, the subtle threats dropped by the traffickers ensured they would comply. The traffickers would stop by every night to make sure the kid was still there. If he were gone, there would be hell to pay. Since the bad guys had gone through the trouble of kidnapping the kid, they would make an attempt to extort some money from his family, on the side. But, the kid had no family and no one from his village had any money to pay. He was screwed. While he wasn't being mistreated by the occupants of the home, the kid was physically shackled, and bound with chains to prevent escape.

The informant had been contacted by one of the men who lived in the home. The men were looking for a way out of the situation they had been forced into. They wanted to just let the poor kid go, but couldn't. They couldn't contact the police, because they were scared the police would not listen to their story and arrest them. In either instance, the bad guys would behead them for losing the captive. They were in between a rock and a hard place.

The occupants agreed to cooperate with the informant, as long as a plan could be devised to where they weren't blamed by the traffickers for the kid's discovery. They indicated they only had one weapon in the home and would obviously not give us any trouble. However, the traffickers were heavily armed at all times. If they were there during a rescue attempt, they would definitely engage and fight. It would get ugly. The bad guys would always arrive unannounced and rarely came during the daytime.

● ● ●

There was a knock at the door, at which point the waiter brought in a spread of rice, bread, beans, and some chicken. Luke warm Pepsi's and Mountain Dew were distributed among the group. The one thing I disliked about eating among the Afghans was the lack of cold drinks. I loved the food. Everything they made was good and I never had a bad meal. But washing it down with a warm Pepsi was a let-down. I initially didn't think you could buy ice in Jalalabad. You can, and there is an ice factory. It's just a luxury that is regarded as not needed, nor required. The vendors will often have an old chest-type freezer they keep their drinks in. Since the drinks were never cold, I don't think any of those freezers actually worked.

When you eat with Afghans, you sit on the small mats or cushions that ring the room. Sometimes, you will just sit on the rug that serves as the carpet. A vinyl tablecloth is then laid out on the floor, in the center of the room. If you are afraid of germs, this is where your fear will begin. Because, as the vinyl tablecloth is laid out, the waiter will be stepping all over it with bare feet. Immediately thereafter, the flat bread, which is the foundation of the meal, is laid on the vinyl mat in front of each patron. Usually there's enough for one per person, placed squarely on the spot where the waiter was just stepping with his grubby feet.

Then, the pot of meat, the pot of beans, and the pot of rice are served. At that point, it's a grab-fest with bare hands, which dig into the pot of rice with a scooping motion, in order to drag a quantity to one's bread. The diners will then tear off a small piece of bread, and use it as a grip for a piece of meat or bit of beans, from the now community pots. This process gets repeated. Hands are flying everywhere with no one observing the rule against double dipping. Rice gets sucked off of fingers like they were lollipops.

You can be strategic, though, if you are efficient about things. The trick is that once the food is laid out, you must strike quickly. Reach in the rice first, and get a big scoop with your bare hand. Drop that on your bread, which will serve as a gringo dinner plate. Then, tear off a big piece of bread and use it to scoop a good first, and final, serving of beans. Drop that on the improvised dinner plate. Repeat the process with the meat. Now you have secured your meal, with the least amount of germ sharing with your Afghan buddies. You can sit back and enjoy your food with a bit of privacy and peace of mind. That is, as long as you don't think about the fact that your edible dinner plate now sits on ground tainted by the waiter's feet. Feet that were recently squatting over a Turkish shitter. Different cultures do different things. I respect that more than anyone. While most will find these arrangements less than acceptable, I adapted.

...On one occasion, I made a huge error in my table manners. You see, I'm left handed. Ambidextrous actually, but I eat with my left hand. In that part of the world, the left hand is what's used to wipe your ass and therefore, considered unclean. But, it's subjective as to when you're focused on it. It's OK to give someone a hug with both hands. OK to cook with two hands. OK to carry bread with two hands. OK to cut vegetables with two hands. NOT OK when you're the first to dig your bare hand into a huge pile of rice, while everyone shrieks in fear, as you realize it's your left hand. The hypocrisy! It's subjective, I'll continue to argue, but it's not good for cultural relations when you're living in their world. Sammy had to constantly remind me to eat with my right hand, before it became habit.

I ate with the Afghans every day in various places with various levels of cleanliness. I only got sick on a few occasions. The cause was mostly from the higher-dollar restaurants and not from eating at someone's home. When I got the runs that weren't going to go away without assistance (i.e., the runs that necessitate being within five meters of a toilet at all times and having your own private roll of toilet paper), I would double down on my doxycycline tablets for a couple of days. That always cured me. No need for further intervention or to go see the medics.

The doxycycline tablets were mandatory issue to all personnel deployed to Afghanistan because of the threat of Malaria. You were supposed to take one every day to ward off this threat. I initially took them as prescribed, but my stomach always felt bad. Doctor Mark told me it was OK to stop taking them. It was a bit of a risk, I suppose, since we had a dozen cases of Malaria on Prison Camp Fenty, during the month of September alone. The cause publicized was that soldiers were not taking their malaria pills. The only good those pills did for me was to end the flow of water from my ass after a bad meal.

● ● ●

The discussion continued, during which the informant provided more information about where the hostage was being held. We decided to have the informant relax and camp out in place, while we went back to the safe house to discuss tactics. The informant had a full belly and a place to sleep, so he was more than content.

In the colonel's office, our options were kicked around. The option of conducting a nighttime raid was discussed, upon notification from the informant that the bad guys had arrived at the location. This scenario would allow for the possibility to arrest the kidnappers. It was quickly scrapped because of the approach into the village. According to the informant, the village was several kilometers off of the paved road. After traveling down a rudimentary dirt road, we would have to dismount and patrol into the village. That distance was roughly six hundred meters, over open terrain with little cover. If the traffickers were at the location, they would have ample time in which to either flee or man defensive positions. The informant advised it would be nearly impossible to describe which house it was, without him going back out to the location and obtaining a GPS coordinate. There were just too many variables stacked against us to conduct a nighttime raid, in unfamiliar territory. We weren't looking for a firefight. We were looking to free an innocent kid from captivity. The nighttime option was ruled out.

The simplified objective of the operation was to safely free the kid, and nothing else. I have learned through hundreds of dope deals that the more complicated you try to get on an op, the greater the chance it has of getting screwed up or not going at all. Keep it simple and it has the best chance of going. Don't get greedy. Through further investigation, we would be able to identify the kidnappers and arrest them at a time of our choosing. In order to prevent retaliation on the three occupants of the home, it would be made to look as if they had no choice but to flee upon our approach. The CI would contact the occupants and advise them to clear on out the back when we got close. Everyone in the village would be able to corroborate the story, that they had no choice but to run.

The decision was made to take the informant with us on the raid. He would be dressed in a police uniform with his face covered, in order to blend with the patrol. That way, he could lead the team direct to the target location. There would be no confusion as to which house it was.

We would adopt this practice on every operation from there on out. Our informant would always be dressed in a police uniform. We would even give him an AK-47 (with no ammo) to carry. This ensured we were not being led into an ambush. If we were, that poor bastard would be the first one shot, or the first one to step on a mine. We never told them this was the plan until we were ready to roll out. That way, if they balked about wearing the uniform and walking point, we would know it was a setup. It would also ensure there was never any doubt as to whether or not we were at the right house.

The U.S. military would frequently raid the wrong house on their operations. *Oops.* I recall one operation the SF unit did. It was hush-hush (secret squirrel, as usual) for those not directly involved. It wasn't my show and didn't concern me. The only reason I knew about it was because some of the EOD techs had been roped into going on the raid. They did rehearsals and trained for two weeks on how they were going to hit the place. After two weeks of training and preparation, the helicopters sat them down in the wrong location. They hit the house strong and initiated the raid. As one of the male occupants was holding up an Afghan government identification card, yelling that he worked for the water department, he was taken down and flex cuffed. Turns out he did work for the local government and not the Taliban.

Oh shit, we're at the wrong house, I'm sure more than one person yelled as their stomachs tightened. Once the slight error was discovered, the team had to run down the street to the correct target location. Upon arrival, it was nothing but crickets. There was nobody at home, or "a dry hole," as the military calls it. Need I explain why? Could it be helicopters landing on a city street, perhaps? Could it be the fact the Americans were raiding the neighbor's house? Could it be because of all the noise and the plethora of U.S. soldiers? If you were a bad guy in the area and didn't haul ass out the back, you deserved to get caught. Or maybe, it was the fact the U.S. military had taken two weeks to plan a simple raid. If someone tells you that a bad guy is at a location, what in the hell would make you think he'd still be there after two weeks? While hitting the wrong place happened frequently to the U.S. military, it happens to law enforcement in the States as well. I'll concede that shit happens.

*...**It's a whole lot different*** looking at a house from street level versus surveillance photos taken by a Predator, no matter how long you study them. We never had that issue because the snitch was always with us. The snitch could immediately be queried if additional information was needed. The snitch knew exactly what house it was, and which door to go through. If the target location was in an area where we could do a drive-by, we'd personally lay eyes on it ahead of time as well. No need for surveillance photos, Predators, or secret maps.*

After finalizing our operational plan, it was time to go see the general. We made a phone call to his assistant, and we were again on our way to the PHQ, to secure our muscle for the raid.

This time, the meeting was more of a social event. We discussed the success of the hash bust and how it went like clockwork. We were in no real hurry because we planned on hitting the place the next morning. We had time to drink tea and hang out on this occasion. When the conversation turned to the business at hand, I told Sammy to have the colonel do the talking. This caught the colonel a bit off guard, because he assumed I would deal with the general. He figured he would be able to relax and let the American ask for the support. He knew it would be harder for the general to turn down the request when it was coming from the gringo. However, it didn't build any of my team's capacity if I was always the one making contacts and asking for favors. The colonel needed to be knockin' taters (redneck slang term for socializing; or physically knocking your fist together with that of another, as a method of greeting) with the players. There would come a day when he would be on his own in dealing with local politics. I, the American, wouldn't always be there as a "show of force." By making the colonel do the talking, it was an additional step in making them self-sufficient.

The colonel did a good job laying out the plan, which was a bit complex due to the terrain and the potential for resistance. Afterward, the general hit the magic doorbell button. Out of thin air, the young, slender lieutenant appeared, and was ready to copy. After a few words from the general, he was off to the races.

Sammy indicated that we needed to roll. Per the general, we needed to go see the colonel over the Criminal Investigations Division (CID). He would be in charge of this operation since it involved a kidnapping.

The CID colonel was a tall, slender man, with gray hair. Our colonel knew him well, from the years they had both spent in police work. We sat in his office explaining the dynamics of the case. After a final round of tea, the meeting was adjourned. It was agreed that all involved would link up at 6:45 a.m. sharp, the next morning. He would have his people ready to do battle. He would also grab a prosecutor to roll with us.

● ● ●

"Equipment check!" I yelled, as we returned to the safe house.

In Pashtu, there were a few *not again* remarks I'm sure, especially from young Agent F, the logistics officer. He knew his equipment room would once again be in shambles.

As the soldier backed the vehicles into ready positions, the rest of the men started getting their gear together. Agent F reluctantly opened the equipment room and I was the first one through the door. After throwing the heap of equipment out into the hallway to be loaded into the vehicles, I walked back outside to find the young soldier staring at me with a gazed look. I interpreted this as him waiting for the go-ahead to top off all the vehicles with fuel. I pointed to the fuel door and made some hand motions, as if I was wrestling an imaginary gas can. He got the picture and went to work.

As the agents assembled by the vehicles for inspection, I began to make my rounds. The first team did better than last time but only two people had rifles.

"They say that two people with rifles are enough," explained Sammy, when my interrogation began.

"Let me guess, no problem right? Sammy, we'll sit here all afternoon and all night until every man has the right equipment. That includes each man bringing a rifle."

Sammy started down the path of explaining that the rifles were crap, but I stopped him before he got too deep. *Just tell them to get their rifles so we can move on.*

Only half of the men had their handheld radios. The rest were safe and sound inside the equipment room, nestled in the charger. After not budging on the issue and after another speech from me about readiness, we were finally prepared to do battle the next morning.

The colonel dismissed the men with instructions to be back at 5:00 a.m. I took this opportunity to address a problem in our logistics procedures that was hampering efforts. What I had come to realize was the colonel kept all of the agents' equipment locked up tight. The only piece of equipment the agents had on them was their Makarov pistol. Anything else needed was to be cleared through the colonel and signed for through Agent F. This isn't the way police officers are supposed to operate. If they had to roll out on something at a moment's notice, they didn't have time to be doing paperwork. They needed the gear to be within arm's reach at all times.

After explaining to the colonel that I wanted the gear issued out to each individual agent and not kept in the cage, I received another lesson about how Afghans do business. The colonel began relaying his thoughts on this request. As he spoke, his voice got louder while his hands were flying with animation. Sammy could only nod his head and say *saysh*, which means yes in Pashtu. Out of breath and with blood pressure off the charts, the colonel rested his case.

Sammy explained that the way the logistics system works is that someone has to sign for all the equipment. That someone was the colonel. If any piece of equipment was found to be missing, broken, damaged, or otherwise unaccounted for, it was on him. Only he would be held responsible. That meant that either he had to pay for the piece of gear out of his own pocket, or possibly face going to jail for losing government equipment. Therefore, he kept everything under lock and key to make sure there were no problems. This was more important to him than the men having easy access to the gear.

It was my opinion that this mentality had been ingrained in the colonel, and some of the others, because they had worked for the Russians when they were younger. The Russians obviously have a different style of management than the Americans do. I wasn't trying to stuff American mentality down his neck and say that I was always right, but this issue was all about commonsense. It was a major safety issue.

I came back with the argument that once the men signed for the equipment in Agent F's book, they were responsible and not him. Just have the men sign for all the gear and hold them accountable. That's what routine inspections are for. If there were any issues, we would deal with them individually.

"One day, you will go back to your family in America. Once you do, I will still be held accountable for this equipment. Who will protect me then, if something is missing?"

I couldn't argue with that statement. The Americans had been in his country for a decade. He had seen people like me come and go. He had probably adapted to dozens of different ways of doing business, due to the constant ingress and egress of military personnel and advisors. I was just one more gringo among the masses.

"Look colonel, you're right. One day I will leave Afghanistan, but it's not going to be anytime soon. Before I go, I'll make sure you're straight on any issues that may cost you money. I'll see to that. But while we're here running operations in the mountains and in areas where we're not welcome, the men will be prepared. That's not negotiable. That's the reason the Americans bought this equipment, so it would be put to use. From now on, it's to be issued out. If something gets damaged or gets lost, that's part of doing battle. We'll get it replaced."

The colonel wasn't buying it, but he knew I always held the trump card. I could easily go to Kabul and raise hell about him with the general in charge of the program. He might find himself being relocated to Kandahar the next day. I would never threaten him with that because he was pivotal to the operation. I just needed him to toughen up a bit and quit being such a pushover. It was time to get a bit gangster to ensure he understood I was serious. The safety of the men was at stake.

"And, the key to the equipment room needs to be available at all times. Agent F should pass his key to the duty agents, when he leaves for the night. If I can't get into that room when I want to, it means the men can't get in there, either. They need to be able to access the equipment in case there's an emergency. The next time I can't get into that room because no one's got the key, I'm going to take the door off its hinges and chop it up into firewood."

That's all it took. He already thought I was crazy so he certainly didn't doubt I would chop up the door—and I would have. The radio chargers were in there with the extra batteries. The extra ammunition was in there. Our frags and RPGs were in there. Should the safe house get attacked, that storage room would serve as the Alamo. The colonel reluctantly agreed to the new procedures, if nothing more than to get me out of his office for the night. The next morning would come early.

Let me say this about the colonel. I respected the fact that he took good care of what had been issued to him. This was like night and day compared to dealing with Iraqis. The Iraqis sold every piece of equipment we gave them. They trashed everything else, and I mean everything. There was no accountability at all. The Afghans were completely the opposite. I had to fight the colonel, just to get him to open up the equipment room. I told the colonel I highly respected him in this regard, but suggested he place a higher priority on readiness. He also needed to put more concern into our mandate of catching bomb makers, versus all other issues.

● ● ●

The alarm went off in my cubicle at 4:00 a.m. I had only been asleep for about three hours. I used my flashlight to get dressed because turning on my light would have illuminated the entire bay. My bumping around had already woken up several of the light sleepers, who gave out a few sighs to express their displeasure. The weather had grown cool at night and I figured it was about time to sport the poncho I had bought. I grabbed my ditch bag and my newly acquired AK-47, and quietly slid out of the building.

...I started carrying an AK-47, *for several reasons. First of all, if I carried my M-4, it had to stay with me at all times. If I got out of my vehicle, it had to go with me. I couldn't leave it in the car. You don't exactly blend with the locals when you're toting an M-4 while wearing man-jams. If I left it in the vehicle and someone stole the car, I would be up shit creek. If you lose your U.S. government-issued weapon, you lose your job. Therefore, it was safer to carry the AK. I could leave it locked in the car if we stopped somewhere, and if it was stolen, it was "no problem" to just buy another one. I even stopped carrying my M-9 pistol. My issued weapons typically remained safely secured on the prison camp, unless my day was spent exclusively on an op.*

I headed out to the car. There was a slight film of ice on the windshield, which I scraped off with a discarded phone card from the floorboard. The Forerunner I drove at the time did not have heat. It was the worst thing that could have happened to me because I can't stand the cold. Out of all the things that could break, it had to be the heater. Why not the power steering? I'd prefer to have to manhandle a steering wheel, over having to freeze my nuts off at four in the morning. Since we had no contract in place to fix our vehicles, I just suffered through the winter.

While shivering, I changed into my man-jams and donned my new poncho at the shipping container. *Man, this thing is stylish. This looks good,* I thought to myself. It was black, with brown fringe that had various designs embroidered in it. I felt like Clint Eastwood. *The blue burqas will barely be able to keep their hands off me when I strut through town. It could cause serious problems.*

I had to honk the horn when I pulled up to the gate, just to get someone to come down from the guard tower. On this particular occasion, the soldiers were obviously sleeping. With headlights shining out of the darkness, you can't tell me no one saw me pull up. Finally, I was allowed to escape.

When I pulled into the safe house, most of the men were already there. The generator was running and the front porch and parking area were both well lit. As I exited the Forerunner, a round of laughter broke out. I looked around to see what was so funny. I couldn't figure it out. Was I dragging something behind my ride? Maybe I had run over a dog and didn't realize it. Grown men were rolling on the ground. There must have been a joke that I missed. As it began to subside, random words in Pashtu would reignite the chaos. I stood there, not knowing whether to laugh with them or say, *What the hell?*

Sammy emerged through the front door of the safe house to see what all the commotion was about. It took him about five seconds to solve the mystery. He froze, as he took a look at me standing there with my new poncho.

Puzzled, he walked over and said as serious as he could be, "Sir, you have a hole in your blanket."

Laughter again erupted—twice the intensity. Grown men had tears in their eyes.

"Yeah, no shit. How do you like my new poncho, and what the hell's so damn funny?" I asked.

"Sir, they're laughing at you because you have a big hole in your blanket. What happened to it?" Sammy inquired.

"This is what we call a poncho, my friend. I had this made in Kabul."

"Sir, you've got to take that off. I will buy you a blanket that has no hole in it," Sammy offered.

"But that's the purpose of the hole; so you can wear it like this. I paid extra for the hole," I tried to explain.

"Sir, they are laughing at you. If you don't take that off, you'll get shame," Sammy insisted.

"I'm not trying to win a fashion contest. I'm trying to keep warm. It's cold out here."

"I'll find you a blanket. But trust me, you've got to take that off."

With that, Sammy helped me out of my Clint Eastwood / Pancho Villa attire. *Where the hell was The Big Chicken?* He was the one to blame for this debacle.

"They all want to know what happened to your blanket. What should I tell them?" asked Sammy.

He was trying to look out for my best interests but didn't know which approach to take. It didn't concern me because I'm not a shame-based person. I don't care what people think about me. No one's going to offend me by laughing or making snide remarks about my clothing.

"Tell them the truth. I paid thirty bucks for the blanket and they charged me an extra twenty to make the hole. In other countries, people wear these things. They're called ponchos."

Sammy explained to the troops about how I had acquired the garment. They had a round of discussions about how those tailors on Campus Eggers ripped us off. *You can buy a blanket for three or four dollars, if you didn't know.*

While they were conferring with one another about my horse blanket, I had a brief laugh to myself. I thought about those three Special Ops guys in the picture The Big Chicken had found. Those three gents roamed around thinking they were the coolest cats in the country wearing those ponchos. But, deductive reasoning tells me they didn't spend much time off the prison camp, nor did they spend much time around Afghans. If they had, they would have been laughed at like me and forced to scrap the attire. They definitely were not doing undercover work. However, they may have indeed been the coolest cats on the campus in the eyes of other Americans.

It made me happy to know the men enjoyed a good laugh at my expense. It was just another incident that bonded us together. A final equipment check ended the festivities. We stopped off in the bazaar to retrieve the informant, who was waiting for us at the curb. We headed out on Highway 1, in the direction of Torkham Gate. Jalalabad was soon in our rear view mirror as the sun slowly started to peek over the horizon. The plan was to meet our ANP brethren at a police substation about thirty minutes outside of town.

● ● ●

We were a bit early for the linkup, which allowed us to stop off at a roadside café for some breakfast. The café had small seating areas built on wooden platforms about two feet off the ground. A quick inspection of the infrastructure didn't leave me with much confidence that it wouldn't crash to the ground, as several burly men climbed aboard. Regardless of my opinion, it held. We had an excellent meal of fresh, warm bread served with a sweet, white, creamy topping. It reminded me of marshmallow crème. Whatever it was, it was delicious. It was times like this that I felt sorry for all of my friends stuck on the prison camp. They were being robbed of the best cultural experience they could ever imagine.

As we pulled up to the ANP substation, there was a convoy of Ford Rangers already assembled and ready to roll out. Men scurried about, loading last minute lots of ammunition and RPG rounds. Gunners manned their PKM machine guns as if they were already underway. Once again, I was impressed with the Afghans. It seemed all they needed was for the overkill of Americans, trying to tell them what to do, to merely step aside and let them handle business.

While the colonel met with the CID commander, our agents with the informant pulled off into a secluded area across the road. The informant was outfitted with his police uniform, a face covering, and was issued his AK-47. There was only one problem. No one brought any boots for the guy.

"Hey sir, can we use those boots in the back?" Sammy asked, while peering into the hatch.

"Those are my boots. How come nobody brought the snitch some boots?" I asked. Sammy shrugged his shoulders as he obviously didn't have the answer to the question.

"Go ahead, but I'd better get them back," I said, knowing full well that if those boots fit, they would become the express property of the snitch. (I never saw those boots again.)

After the prosecutor was brought up to speed on the case, it was a race down the driveway and into traffic, with blue lights flashing, sirens wailing, and heavily armed Afghans ready to do battle. Our vehicles led the way, for only key personnel involved knew where we were going. Our only stop would be at the Achin District Center, where we would notify the district chief and pick up additional personnel. Less than an hour later, we had completed the final link up and were now heavy an additional five troops and one Ford Ranger.

The paved road was only beneath us for about ten minutes thereafter. When we hit the dirt road, the game was on. There was no slowing down other than to endure that initial bump from the absence of asphalt.

"Everybody wants to go to heaven…"

"But nobody wants to go now…"

After Sammy and I knocked taters (knocking fists together) we yelled, "*Fuck* it, let's go!"

The first few kilometers of the road saw few homes. The dust was thick upon us, as we were soon jockeyed to the third position in the convoy. The local officers knew how to navigate the road and were familiar with where it ended. They would lead us the rest of the way. A discussion ensued as to why I wouldn't roll up the windows or close the sunroof. The dust was unbearable.

"Sammy, if the windows are up, no one will be able to figure out from which direction we're being shot at. And how can you three shoot back if you've got to take the time to roll the windows down? These windows aren't bulletproof anyway, and a little dust never killed anyone. Matter of fact, get your narrow ass in the sunroof and spot."

...The dust was bad, but my insistence on having the windows down was a remnant from working the road as a rookie cop. If you drive through a neighborhood with the windows up and the air conditioner on, with a stereo and a police radio making noise, you can't hear a gunshot go off a few meters away. I compare it to being inside an MRAP. Your senses of hearing and smell are diminished. Therefore, we would endure the Afghan dust until we reached the staging area. With Sammy posted through the sunroof, we had an improvised up gunner. He could also spot any freshly moved dirt and potentially prevent us from hitting a mine. Watching him, he reminded me of those two kids in the bird's nest on the movie "Titanic." Hopefully, he would do a better job than they did.

The convoy stopped on a couple of occasions for the commander to deploy dismounted personnel on key ridgelines. They would provide overwatch on the route, in order to prevent the bad guys from digging bombs in the road and surprising us when we departed. This was smart, and was a common practice when I worked with the Afghans. This is not something the Americans would typically do. They felt it was too dangerous. Meanwhile, it allowed the bad guys free reign to do whatever they wanted to the roadway, and allowed them to lie in wait for the convoy to come back through.

On several occasions, I had to kick the Forerunner in 4-wheel drive to get up some of the inclines. As always, it performed flawlessly. I would come to believe that Toyota Forerunners were the toughest vehicles ever made. Dodging huge rocks, holes, and sand traps made driving a constant evaluation process. Go left, go right, take that fork, slow down, speed up. A wrong or unlucky decision could end with a bang. I drove with my right hand and held onto the "oh shit" bar, with my left.

I thought about the MRAPs the military had. There was no physical way they could travel where we were. They were too wide and too big.

They had recently come out with the new MRAP which was supposed to be designed for Afghanistan. It was just as wide as the other MRAPs, only shorter. That made no sense. How was it supposed to traverse the narrow mountain passageways any better? The answer is that it wasn't and it couldn't. It was just another concept the military bought off on due to creative marketing practices and lobbying by the defense industry. The military didn't need a new version of the MRAP. It was just another way for the defense contractors to make a buck. I'm sure whoever supported its fielding will certainly have a nice job waiting for them when they retire from the service. MRAPs were about a million dollars each by the time they were outfitted. My Forerunner cost about six grand and could go anywhere in Afghanistan.

...On one occasion in Iraq, I attended a briefing on Afghanistan by a Marine lieutenant colonel. He acknowledged that most would not agree with the picture on his first slide. It was a picture of several young Marines riding in the back of an ANP Ford Ranger. One had on a Kevlar helmet, but no body armor. One had on body armor, but no Kevlar. One had on a USMC eight-point soft cover, but no body armor. Everyone in the room silently shrieked because the Marines were not "properly" outfitted with body armor, Kevlar, safety glasses, gloves, etc. Plus, they weren't riding in an MRAP! Yikes! If they got hurt, someone would get a bad fitness report for sure! The lieutenant colonel explained. He told his Marines that their mission was to support the Afghan police with their operations. That's what they had been ordered to do. It was exactly what they would do, until ordered otherwise. That meant humping mountains and treacherous terrain. They were to keep up with the Afghans who were accustomed to humping mountains. He told them they could wear anything they thought they needed to, in order to accomplish the mission. They would only be questioned if they couldn't keep up with the Afghans. The lieutenant colonel went on to explain to the reluctant faces in the room, that it wasn't their mission to stay safe. If it had been their mission to stay safe, he would have simply kept them at Camp Lejeune and not deployed. It would have accomplished that particular mission. The mission was to support the Afghans, and from his perspective, everything else was secondary. This gentleman was one of the smartest individuals I've ever met. However, I don't think he'll be successful in the U.S. government because he's efficient, has commonsense, and is mission oriented. Plus, he's got the balls to buck the system. However, if a private company needs a chief executive officer, he'll make them millions.

The road suddenly ended at a circular plot of flat ground. To the immediate right, there was a ridgeline. It offered an excellent place from which the bad guys could engage us down below. The convoy commander immediately ordered three men with a PKM to deploy on this ridgeline and provide overwatch. They began the long hump up the hill, toting weapons and extra ammo for the gun. Good call on his part. At least we wouldn't have any surprises waiting on us when we got back.

To the left, there was a huge open area on the valley floor. Approximately six hundred meters across this clearing stood a small mountain. At the far, top left of the mountain were a few structures. Near these structures, we saw a large group of villagers who had gathered to see what we were up to. The snitch pointed and advised the target area was past where the villagers were standing. It was going to be a hump getting up there and everyone knew we were coming.

There was no time wasted. As soon as we stopped, we were patrolling across the open area. While the main element patrolled in a ranger file (line) to the right of the clearing, the convoy commander sent a squad of men on our left flank in case we got hit. The snitch led the way as planned. Agent S was next in line, with Sammy and me following. The old prosecutor, who I estimated to be around seventy years old, brought up the rear. Several CID agents, who were dressed in suits and armed with AK-47s, provided security for the prosecutor. Not many seventy-year-old Westerners could hump mountains like that old man was doing. I grew up going to church, religiously (no pun intended). When you think of a wise man, the image of this old prosecutor is what will pop into your head. As we reached the location on the ridgeline where we initially saw the villagers gathering, there was quite a crowd of people. They weren't happy about our presence and the comments began to fly.

"Why are you here? We've done nothing wrong! What do you want? Leave us alone!"

It was very similar to when police roll through the projects back in the States. Nobody wants the police rolling through screwing with them, and nobody wanted us there in the village that day. This was probably the only time that government troops had been in Narghosa.

I was glad no one knew I was American. The people were merely reacting to the presence of the local police and not the invading American crusaders. This had nothing to do with religion or international politics. People were pissed off at the police for hassling them, just like in the States. (People always have something bad to say about the police and never give them enough credit for their efforts. That is, of course, until they need the police because some nut-job is shooting up the place.)

It took us about thirty minutes of a half-walk and half-run pace in order to reach the target house. So far, the mission was going well. We had successfully reached the objective without getting engaged. There had been only a few snide remarks in opposition.

The informant entered the qalat through a darkened passageway. We were right on his heels. There were doors to the left and doors to the right. The informant kept walking forward until the passageway opened up into a small courtyard. The informant pointed upwards at the mud wall in front of us. I didn't see anything. It looked like it was just the wall. There were no stairs or steps.

The informant pulled over a long wooden plank and leaned it up against the two-story mud wall. He told Sammy to use it as a ladder. The young kid was reportedly, in a room built into the wall.

I helped Sammy scamper up the plank and then Agent S helped me up the makeshift ladder. When I reached the top, Sammy and I discovered there was indeed a room built into the far side of the wall. A white plastic curtain was draped down from a small overhang, obscuring our view of the inside. As I covered Sammy, he walked over and abruptly pulled the curtain back. I scanned for threats but immediately lowered my weapon.

A young Pakistani was lying on a makeshift bed with a fuzzy blanket over him. All we could see was a chain coming from underneath the covers, which was locked to a post. The young man began to sit up with a look of fear on his face. I quickly deduced why. Sammy had my good camera hanging from around his neck. I had an AK-47 that had briefly been pointed in the poor kid's direction. He didn't know us. He had never seen us before. All he knew was that one guy had an expensive camera and one guy had a rifle. I was the executioner and Sammy was the camera man, had to be the thoughts going through his head. He was about to get greased.

"Tell him who we are, Sammy. He's scared."

In Pashtu, "Relax my friend. We're the police. We're here to take you home," Sammy said proudly, as he began snapping pictures.

Agent S climbed up the makeshift ladder and asked the young man about how long he had been tied up. "Two months and four days," was the immediate response. He knew exactly how long he had been chained to that post.

The small room was soon crowded as the CID agents, the prosecutor, and several other officers climbed the plank to inspect the goods. The CID agents began their paperwork while Sammy and I made sure we had good photos of the scene. The effort then turned to freeing the kid, known as Dawood, from the chains. I only blamed myself, but no one had thought to bring bolt cutters. The informant had told us he was tied to a chain, but we hadn't thought that far ahead.

Dawood had a heavy-duty set of shackles on his feet which were secured by a large padlock. There was no way we were getting those off. The weak link was the smaller lock which held the chain together around the post. As the old prosecutor looked on, one of the young ANP officers beat the hell out of the lock with a set of pliers and a huge stone. After a good ten minutes, the lock finally gave way and broke.

I insisted we all take a quick trophy shot with our victim. With Dawood in the middle, and several agents holding up his chains, we took some photos that turned out good. The old prosecutor was right there in the middle of the photo shoot.

The only way for us to get Dawood back to the vehicles was to carry him. It was a long hump down the mountain and across the clearing, but we didn't have any other choice. Teamwork got Dawood safely down the plank serving as a ladder. Agent S heaved Dawood over his shoulder in a fireman's carry. Another one of our agents carried the length of the chain. We walked back through the dark passageway and out of the qalat. All of the young Afghan officers who had been fighting boredom while standing watch on the perimeter, were now able to share in the excitement. They had come for a reason. There was no "dry hole" here.

The entire village had now amassed to see what the commotion was all about. As the agents strolled past them with a young kid shackled at the feet, the mood changed. Once the crowd of people had a grasp on what had happened, they began to thank the officers for their efforts. The same elderly women who had cursed us on our way in were now crying for young Dawood. While it's hard to keep a secret in a small village, it didn't seem to me that anyone was in on it. The villagers seemed bewildered that the kid had been there. No one knew who he was. Where was his family from? No one had any idea. This exchange was a good sign. It meant that the people were with us on this case. The only ones to be offended were the traffickers.

While the terrain was rugged, everyone wanted a turn at carrying our guy. I didn't last long and had to hand him off after a few minutes.

He definitely hadn't missed any meals while in captivity. It was not until we got back to the vehicle staging area that reality set in for Dawood. Once he saw all of the police vehicles and the overall mass of officers, I think he finally realized he was going to be free. He shed a few tears and tried to kiss everyone's hand. A smile finally returned to his face as we took some photographs with him before departing.

Our journey back to the paved road was met only with curiosity from onlookers. We picked up our overwatch personnel, who had kept us safe from ambush. A sigh of relief came from Sammy and me as the mission had been accomplished. Little did we know, we would eventually return to Narghosa under circumstances with a bit more conflict.

● ● ●

On the way back to Jalalabad, Agent M got a call from an informant. The snitch said he had a couple of Taliban fighters with him, who had information on a higher-level Taliban commander we were looking for. They specifically wanted to talk with "the American" and asked us to meet at a restaurant close to J-Bad. We were hungry anyway, and decided to go meet the gentlemen. When we got close, Agent M told me to stop the car a few buildings down. Sammy and I would survey the scene while Agent M went inside to confirm that it wasn't a trap. This particular snitch was reliable but we didn't know the crew he was bringing. Therefore, commonsense applied. We would assume it was a setup until proven otherwise, because they knew the gringo was coming.

About ten minutes later, Agent M called Sammy and gave him the code word that everything was straight. We locked the Forerunner and made our way to the restaurant. We walked in and were led to a darkened room at the back of the building. Either the lights didn't work, or they purposely wanted it kept dark. Regardless, as we walked in, there was an eerie feeling as the snitch introduced me to the two Taliban fighters. As I shook their hands, I could sense their distrust and apprehension about being in the same room with me. For a few moments, I really thought they were about to start shooting, or maybe even just bolt out the door and run. The silent tension began to escalate. Different courses of action began spinning through my head. We could take them if they decided to break bad, but if someone pulled the pin on a grenade, we would all die in that small room. I rolled the dice.

I decided to sit down and give them the higher ground. I wanted them to feel a sense of domination and control for a moment. Although not a good move tactically, I hoped it would lessen their perception that I was a threat, or a spy sent to kill them, by exhibiting confidence in them first. The gamble worked and cleared the tension from the air. They followed suit and the conversation began.

With Sammy translating, we learned of their story and what they wanted. The one gentleman was the equivalent of a squad leader and the second gentleman was merely a soldier. The leader explained that he and his group had not wanted to fight for quite some time. However, it was their only means of income and it was difficult to just walk away. Through a series of meetings and conversations among themselves, they had decided to surrender and reconcile with the Afghan government. The reconciliation program, at the time, purported that if Taliban fighters would reconcile, they would be given land, a small amount of money, and assistance in finding a job. They were required to bring in their weapons at the time of reconciliation.

The group had traveled to the reconciliation office in Jalalabad and given up their weapons. They filled out some paperwork and were told they would be notified once their case had been approved in Kabul. Thank you and goodbye. They found themselves outside the office without their weapons and with no immediate assistance. An Afghan Border Police colonel, who happened to be at the office, gave them money out of his own pocket for food and the trip home. Weeks had passed since they reconciled. Their calls to the office, about the status of their cases, repeatedly resulted in the mere response that their papers were in Kabul.

They had already taken a rash of shit from their Taliban commander about what happened to their weapons. They told him they had to abandon their weapons and run during an engagement with an American convoy. It was a weak story he really didn't believe. The other thing was, there were about a hundred other Taliban fighters who were watching to see how their cases were handled. If the government helped this first crew, there would be Taliban members reconciling in droves. But so far, the only thing they could tell was that these guys lost their weapons and wasted their time in Jalalabad. It was perceived to be just another lie from the government in Kabul and the Americans.

They had talked with the informant and expressed an interest in working for us. They were looking for money to feed their families. They knew several higher-level commanders, they knew the locations of caches, and could provide real-time information on pending attacks. My perception of these guys was that they were just poor people, trying to survive. They were really indifferent toward the Americans. They had no choice other than to go along with the program—a program they weren't motivated about being a part of.

As the food was served, the mood lightened. No matter how much distrust you have for a person, it diminishes somewhat when you're passing plates of food back and forth. By the end of the meeting, we were all just a bunch of good old boys talking about women, who had the best cell phone, and about where you could get the cheapest Pakistani hooch. As the meeting was concluding, I thought about the quagmire they would probably find themselves in.

"If you gents have to participate in any attacks to stay in the good graces of your commander, I need you to do me a favor..."

I paused to allow the translation and then continued.

"I need you to aim high."

Another pause for the translation resulted in puzzled looks.

"These young soldiers are no different than you guys. They're mostly poor kids trying to make a living. They're sent here by the rich and don't have a choice. I can't help you guys if I know you're still directly involved in the killing."

...When you're dealing with informants, *you need to realize that they may have to be present during heinous acts, in order to maintain their status and provide you with accurate information. Nobody will trust a person who suddenly refuses to participate in the cause. The goal is to figure out a way to lessen their direct involvement and culpability in the plots. In this particular case, if these guys went along with the program and fired a few rounds high and to the right, I could live with that. They had no motivation about killing Americans, anyway.*

We left the lunch meeting with a new insight to the fraud and failure of the reconciliation program. I would surmise that funding had been pumped into the program, and that said funds had been siphoned off at the highest of levels in Kabul. As usual, we Americans weren't smart enough to ensure our money was spent where it was intended.

We had already been in negotiations with other crews and facets within the Taliban, to surrender and reconcile. We would eventually be responsible for over one hundred fighters coming to Jalalabad and turning over their weapons. The problem was that the program was just a house of cards. It was merely propaganda. No Taliban fighter I know ever received any assistance after coming in and reconciling. Our reputation began to suffer because of this, to the point we decided to scale back our efforts in encouraging people to reconcile. We had to, in order to maintain our credibility among the population. Had the program been real, who knows what would have happened.

CHAPTER 5

Islam and Beer

I received word from Jim that I had been assigned a new interpreter from the States. An e-mail also came from the company which fielded interpreters, known as Mission Essential. The new interpreter was standing by in Bagram, just waiting for me to give them the green light to send him to Jalalabad.

Let me explain how the interpreter system works in Afghanistan. There are many locally hired interpreters from the community. They are locals, but are referred to formally as *local nationals* (LN). Using the term local national was like referring to me as a contractor. It's somewhat derogatory in nature. When used, the speaker's perspective is that of an elitist. Regardless, it's how the Americans refer to the locals. The perception was, and is, that local nationals are not to be trusted.

A local national interpreter makes between $600 U.S. dollars and $900 U.S. dollars per month, depending on his assignment and length of employment. They have no benefits, no health insurance, no dental insurance, and no real vacation time. If they take a day off, they don't get paid. They walk patrol with the military in even the most dangerous areas, sometimes on a daily basis. They do not have security clearances so they can't access a lot of the buildings on the prison camps. They can't get a clearance because they are not United States citizens. They have to get searched every time they come and go from the prison camps as if they are criminals. There is more, but basically the Americans treat them like third class citizens.

The benefits of having a local interpreter are numerous. First of all, they are obviously fluent in the local language and can jive with the locals. Even if their English skills are lacking, they still understand most of what you're asking them to do. They know the terrain because they grew up there. They know the players, bad guys, politicians, police officers, soldiers, etc. Even more important, they know the history between all of these actors. They know why a certain general doesn't like the governor, etc. They can tell you the real story about things. Most of the local interpreters have been working for the Americans since the war began. They have been doing the job for ten years without a break. They have walked more patrols and have been in more combat than any one particular U.S. service member—and without any recognition or medals. They know how to shoot an AK-47, an RPG-7, and how to throw a grenade. They know how to survive. They've been doing so their entire lives. They lived through the Taliban regime or were forced to flee to Pakistan and live in refugee camps. They are tough and are invaluable.

But, we don't recognize this nor treat them with the respect they deserve. We subject the Afghans to the same prejudice that we inflicted upon the local Vietnamese who worked for us during that war, prior to us abruptly abandoning them to a certain death.

The only downfall of having a local interpreter is that because they don't have a clearance, they can't access the Secure Internet Protocol Router (SIPR) network. Therefore, if something is on SIPR, they can't sit down and translate it for you. That's their only limitation but it's really not that important anyway.

Let's contrast a local national interpreter with an interpreter from the States. The U.S. hire possesses a SECRET security clearance. Therefore, they can have a SIPR account and are privy to the information therein. They make around $200,000 U.S. dollars per year in comparison to the $10,000 U.S. dollars per year we pay a local guy. Most of the U.S. hired interpreters are older people. They're Afghans who fled to the United States; ten, twenty, even thirty years ago. They've been living the good life in the States and some don't even remember being in Afghanistan. They've missed several regime changes and really have no idea what the hell has been going on inside the country. The extent of their knowledge on Afghanistan is that they have relatives who sometimes call and ask them to send money. That's it.

They don't know the terrain, the players, nor the history. And here's the worse part; most of them can't speak Pashtu! They barely squeaked by a basic language test to get the job. Remember, if they don't pass the test, their company can't begin to bill the U.S. government. Therefore, they will almost always pass the test, because it's all about the money. Being that they are older, they are in terrible physical shape. They cannot hump mountains while wearing body armor. They can barely hump over to the chow hall without breaking a sweat. They are afraid of their own shadows, because they have never had to face the danger of combat. Basically, you can't take them off the prison camp because you risk giving them a heart attack or a stroke.

...In a nutshell, about 99% of the interpreters hired from the United States are as useless as tits on a boar hog. They just sit around and soak up your tax dollars to the tune of $200,000 U.S. dollars per year, or more. I wouldn't trade one local guy for ten U.S. hires.

Regardless, I wanted to give the new interpreter a very warm welcome to our team. I was hoping he would be tough and able to handle himself. We were a specialized unit that took more risks than anyone around. His position would be much more dangerous, when compared to the job assignments of the people he was hired with. I was praying this guy would be a true soldier.

I secured him a room through the first sergeant. Sammy and I gave it a good scrub down and even stole a couple of pieces of furniture to go in it. It wasn't much, but I believe in taking care of my people. Once I had everything in place, I sent an e-mail to Mission Essential and instructed them to send him on. A few days later, he arrived on the airfield.

I had already given him a cover name, based upon his last name. Nothing fancy, but we would call him Mr. G. All of the interpreters took on an alias or cover name for security purposes. In reality, only the locals needed cover names. The interpreters from the States didn't have anything to worry about. However, all of the intelligence briefings they received made them afraid to even use their first names. Remember, fear makes money.

Mr. G was not at all what I was hoping for—more of what I had expected. He was about five-foot-six, with a big beer belly. He had to be pushing fifty years old. Just by appearance alone, it was obvious this guy was not going to be able to hump mountains. Sammy and I introduced ourselves and loaded all of his gear into the Forerunner. We got him moved into his cubicle at which point I suggested we get some chow.

During chow, Mr. G exhibited a timid personality. This wasn't his true self, and was only due to the new environment he had found himself a part of. Later on, once he got comfortable with the prison camp, he would become quite the socialite. Mr. G explained he had lived in Kabul as a child, during the Russian occupation. At some point during this time, he made his way to Peshawar, Pakistan. Upon arrival, he was flat broke. He came up with a scheme to make money by selling small Pakistani flags he had made by hand. Using an old bicycle as a storefront, he sold every flag he made. Soon, with a pocket full of rupees, he made his way over to India. There, he hooked up with a bunch of Afghans who were living off some type of U.S. government assistance program. He and the group of Afghans survived anyway they could, while waiting for immigration paperwork to clear them to the United States. Frequently, they would capture a cow they found roaming the streets. According to Mr. G, there were cows everywhere because the Indians don't eat them. (Cows are sacred to Hindus and are protected in India.) They would barbecue the cow and enjoy the festivities. During one of these barbecues, the Indian police came and raided the house they were in. Through a quick negotiation that involved some bribery, the police agreed to only arrest the homeowner. Everyone else just had to get the hell out of there. His stories were very interesting and I would always enjoy talking with him.

On the serious side, Mr. G had never fired a gun. He had been a truck driver for many years in the States and had been running a sign shop until the economy tanked. He knew nothing of combat, law enforcement, or basic survival skills. The only thing he knew was what they told him during the CONUS Replacement Center (CRC) process at Fort Benning. I quickly came to the conclusion that I could not take Mr. G out on any operations. It wasn't fair to him or his family because it might get him killed. Even more important, it wasn't fair to the rest of my team. I would keep Mr. G safe and sound on the prison camp and use him however I could. I would include him in any festivities at the safe house, but that would be the extent of his travels outside the wire.

We had already planned a big barbecue at the safe house to celebrate some of our recent accomplishments. This would be a good time to introduce Mr. G to the team and give him a feel for what we had going on. Sammy and I procured a few sets of man-jams for Mr. G, from our shipping container. As he tried on the man-jams, he began to exhibit his true personality.

"Do you have any other colors? I really don't like this color."

It was amazing. He bitched about the color. Then he complained about the way they fit. You would have thought he was a woman trying on clothes at the mall.

"You're not here to win a fashion contest," was the most docile response I could come up with.

"Some other interpreters told me there was a tailor on the other side of the airfield. Do we have time to go see if he can make me some?"

"Sure, on Friday when we don't have anything going on. Knock yourself out. But those will do for a barbecue."

● ● ●

The next morning, we loaded up the Forerunner with bottled water and a few cases of Coke I had stolen for the party. I had given Mr. G a thorough briefing on how we travel, what to do in case we got hit, how to call for help, and what to bring. I instructed him to not bring his entire wallet, for good reason. If we were captured, we didn't need the bad guys having all of our documents. The home address clearly noted on a driver's license was my main concern. I asked Mr. G to only bring his military identification card. That's all we needed to enter and exit the prison camp. Other than that, we didn't need ID while out in the community. Obviously, I told him to bring his cell phone and to keep that on his person at all times. Pack light. Bring a bag with some snacks and water in case the day turned long. As Mr. G loaded his bag into the Forerunner, I decided to see how well he could follow instructions. At some point, his life may depend upon how well he can accomplish the task.

"OK, equipment check. ID cards," I said, as I pulled mine out.

"Got it," Mr. G replied, as he pulled his military ID card from a wallet about six inches thick.

"My friend, don't bring your wallet when we roll out. There's no need to carry all of that around. Cell phone?"

"Oh, I left it in my room."

The detailed instructions I had given Mr. G the day before hadn't set in. I sent him back to his hooch to get rid of his wallet, most of the unnecessary contents of his bag, and to pick up his government-issued cell phone. We changed into our man-jams at the shipping container and were ready to go.

As we were about to roll out the gates of the prison camp, he spoke up and asked, "Isn't there a safer way for us to do this?"

"What do you mean? To do what?"

"Well, couldn't we have a bunch of your guys come here and escort us to the safe house? Wouldn't that be a lot safer?"

I was astonished. No, I was pissed. I told him it was the way we did business. If he didn't like it or didn't feel safe, just let me know. As soon as we got back, I would ensure he was transferred to a much safer position, with no hard feelings at all. There was nothing professional in the one-way conversation we had en route. I had enough problems dealing with the military, my company, and every other whiner about how we did business. I would be damned if a member of my own team would question our methods. As we pulled into the safe house, there was no mistaking where he stood with me at that moment. I would later apologize and elaborate on why it struck a nerve with me.

Sammy was there and began to introduce him to the team. Everyone welcomed Mr. G with open arms. They were all very excited to talk with him and find out where he was from, about the United States, and about his family. He was an instant celebrity among the team members. This was a good thing because it would allow Mr. G to brush up on his language skills. His Dari was fine but he could barely speak Pashtu.

I had given Sammy $300 U.S. dollars the day before, in order to buy the sheep and the vegetables. Not long after we got there, the guest of honor arrived in the hatch area of one of our Forerunners. The sheep was pulled out and allowed to roam the grounds as the barbecue pit was prepared. *Might as well get some last minute exercise because you're about to be dinner*, I thought.

The agents who purchased the sheep had also commissioned a butcher for the party. Having the butcher there wasn't good for operational security, but the barbecue took precedence. They knew him and assured me he was a good guy. Besides, as far as the butcher knew, we were a non-governmental organization (NGO) of some type.

The most important ingredient of any barbecue was about to be discussed and subsequently acquired.

"Sammy, how much do we need for the beer?"

With that, Sammy and I went to figuring on how much beer was required. After a bit of calculation, I handed him enough money to ensure no man went thirsty.

"And get a block of ice, too," I yelled, as he and Agent S were driving out the gate.

Thirty minutes later, they returned with a trunk full of Heineken and a big, long block of ice. Soon, most everyone on the team was enjoying a nice cold one.

...Let me elaborate *on this issue for just a minute. Here's a lesson in culture, especially directed to the genius who dreamed up General Order Number 1, so pay close attention. Christians drink. Jews drink. Believe it or not, Muslims drink, too. Surprise, asshole! That's reality. There is beer and* *liquor in Afghanistan. Yes, it's technically illegal in the country. But much like General Order Number 1, it's a law that most follow only in theory in order to maintain outside perceptions. Most of the agents (Muslims) on my team would openly partake in the drinking. There were a few who did not drink openly, but I would catch them sneaking a can or two for the privacy of their own homes. Those were my hypocrites, but it's up to them to guide their reputations. I only had one out of fourteen who truly did not drink. Coincidentally, he happened to be the only useless member of our unit.*

You could usually buy two different brands of beer in Jalalabad. It was either Heineken or Amsterdam. Amsterdam only came in a forty-ounce can. If you drank one Amsterdam, you were drunk. I didn't like the taste but you couldn't afford to be picky considering the environment. They were sold in some of the shops at the bazaar, in a very low-key way. The shop owner would put the beer in a box or a trash bag so that everything was kept underground.

Occasionally, our brothers from the Afghan National Police would do a sweep through the shops and seize all the beer. Now this was very amusing. Some of the same people we were drinking with on Thursday and Friday nights were involved in the raids. Once they seized the beer, what did they do with it? They drank it! That's exactly what they did with it. It was hypocrisy at its finest in order to maintain the perceptions of Islam. It's the equivalent of a Baptist preacher speaking out against adultery, all the while he's tagging the deacon's wife on Wednesday nights after church. Or, the Catholic priest counseling couples on marriage, all the while he's sexually abusing a five-year-old boy.

After the hired butcher erected the barbecue pit with some loose bricks, it was time to kill the sheep. Since it was Mr. G's first time hanging out with us, we offered for him to do the honors. He quickly declined. Thirty years of living in the States had depleted his primal instincts. Meat no longer walked around on four legs. It came in a nice, tidy package that was shrink-wrapped in the meat section of the grocery store. I'm sure his thoughts wandered back to his days of poaching cattle in India, but now he didn't even have the stomach to kill a sheep. Besides, he might get blood on his man-jams. That would leave a stain.

Agent N pinned the sheep to the ground and the butcher did the honors. He slit the sheep's throat with a large butcher knife. Blood began to spew from the sheep's throat, as if a hydraulic line had been cut. Several seconds later, the spewing turned to a slow leak. After a bit of kicking and wriggling with a few gurgling noises, the sheep was deceased. The butcher cut the head off completely and began to dress our meal. We offered to help him, but he said that he didn't need any assistance.

Focus turned back to the conversation at hand, and I decided it was time for me to break out my backpack. The backpack was armed with about thirty of the small, plastic bottles of liquor, like you are served on a commercial flight. I had a variety of rum and whiskey. I had a stack of the red plastic party cups, along with straws, lime juice, and lemon juice—all of the necessary ingredients to make some good old American mixed drinks. It was courtesy of my wife back home, whom I had talked into smuggling at least five bottles per care package she sent. She had sent some crackers and a can of cheese as well. Oh, and a box of Twinkies. All were good appetizers for the crew while we awaited the main course.

Aside from a few differences, it was the same as a typical barbecue in the States. We had a grill. Our grill was made of loose bricks rather than having been purchased at Wal-Mart. We had beer. We had ice. We had Coca-Cola. We had liquor. We had good conversation. We had friendship. We laughed and we joked. However, we lacked a vital element. I love my Afghan friends, but they just don't understand what they're missing by not having a bunch of drunken chicks at a party. As I mixed some drinks, I caught my *non-drinkers* stealing a couple of the bottles for later.

By the time the sheep was ready to go, most of us were drunk. We ate like kings on the freshest vegetables, bread, and meat. By the end of the meal, I couldn't stomach another bite. The colonel elected to say a few words to the team and I followed his lead. Afterward, the festivities began to wind down.

…This should be a lesson for the leadership of the U.S. military on building camaraderie. How many people in the States throw a barbecue or a party and don't allow beer? Not many. So, why would you ban your people in Afghanistan from building camaraderie and true relationships with the people they are supposed to be winning over? Sure, they eat meals at "key leadership engagements." The problem is that it's all about formality, smoke, and mirrors. In reality, you don't build any trust or camaraderie by saying a few politically correct words (via an interpreter) you've practiced and rehearsed. But, if you drink with people, you don't need an interpreter. Drunks speak an international language understood by all other drunks. No need for an interpreter. Our barbecues boosted motivation, built trust, and created a cohesive unit. And for the record, nobody ever got clubbed to death with a SAW or an M-16.

The barbecue was a success and after sobering up, it was time for Mr. G and I to return to prison. We said our goodbyes and rolled out under the cover of darkness. As we approached the gates of the prison camp, I told Mr. G to light a couple of cigarettes to mask the smell of Jack Daniel's in the vehicle. The crew working the gate that night gave us no hassle. Soon, we were both sleeping it off in our cubicles.

● ● ●

When Sammy arrived the next day, the three of us hit the chow hall. Now, Mr. G had confided in me that he wasn't exactly a Muslim anymore. In all outward perceptions he was a Muslim. However, he had been married to a Mexican girl in the States and had been going to a Catholic church for years. His kids were raised Catholic. We both agreed it would be much better if he and I kept that to ourselves, and let everyone else think he was a good Muslim. Sammy would slowly pick apart his cover.

"Don't eat that! It's pork!" Sammy yelled, as Mr. G scooped a couple of sausage links onto his plate.

"Oh, thanks, I thought it was turkey," Mr. G replied, realizing he wouldn't be enjoying any pork sausage or bacon in the present company.

As we sat and ate breakfast, Sammy advised, "Mr. G, I'm going to have to update you."

Sammy meant that he needed to get Mr. G back in line with Islam. Mr. G had been gone too long. I laughed to myself as Sammy began to preach. I was laughing, because yesterday Sammy was as drunk as I was. That's not exactly allowed under Islam, the last time I checked. No one was judging Sammy, but now he was on Mr. G for attempting to eat that sausage. He was the pot calling the kettle black.

This conversation began an almost daily bickering between the two of them. Mr. G was no idiot to the teachings of Islam and would antagonize Sammy to wit's end. At first, it made me laugh until I hurt. But as the days went by, it became annoying. Mr. G gained more and more confidence as he got to know us and would hit back at Sammy. Sammy was serious about the subject while Mr. G found it amusing. This began to create friction which was unproductive.

One day, when it was just Sammy and me in the car, he started preaching. He was preaching in general, but was also specifically referencing Mr. G and his indiscretions. Like any good Muslim, after mentioning the word *Mohammad*, it was immediately followed by, *Peace be upon him.* After hearing the words *Peace be upon him* about three hundred times, I was at the end of my rope.

"Damn it Sammy, that's enough of the religious bullshit. I'm tired of hearing about Mohammad and everything else. I'm not Muslim. I'm not even a Christian. I respect the fact that you're Muslim and I respect your beliefs. But you've got to stop trying to convert me, and stop preaching to Mr. G. You and I are the best of friends. I'd die for you. So stop letting religion come between us. I don't care if you're Muslim, Jewish, Christian, or Buddhist. You'll always be my friend. If everybody thought like that, the Americans wouldn't be here fighting with the Taliban…"

Sammy's icy stare abruptly stopped my Napoleonic rant. Maybe I crossed the line by using Mohammad's name in vain. No, I'm certain I crossed the line. I am ashamed of those remarks to this day, and the temporary lack of respect I exhibited. But Sammy was like a brother. You don't have to watch what you say or be politically correct around your brother. We had that bond so I could say what was on my mind, and I did. That type of bond is what the U.S. military lacks between them and their Afghan counterparts.

From that point on, there were a few snide comments between Sammy and Mr. G every now and then, but the religious war had ended. Sammy is a smart person. He obviously realized at some point Mr. G wasn't really Muslim, and therefore, was beyond being "updated."

…*Religion truly divides.* *For many years, I believed that illegal drugs were the root cause of all evil, due to my line of work. I was wrong. Religion is the root cause of evil, war, violence, and atrocity. The silly part is that the warring factions (Muslims vs. Christians and Jews) all believe in the same god, even though many are ignorant to the fact or just don't want to admit it. We have been killing one another because our methods of living and worshipping (again, to the same god) don't perfectly correspond. Think about this for a minute; if one truly believed in God (Allah), one would not wage war on a fellow believer, no matter what external differences existed on earth. It defies logic to oppose one another on earth, if you truly believe there is an after-life, lasting for eternity. If you truly believe, then you have to concede that Christians, Jews, and Muslims will arrive through the same gates in heaven (paradise). Christians, Jews, and Muslims will all find themselves standing shoulder to shoulder, awaiting judgment by the same court.*

Mr. G explained to me that aside from his immediate family, every single relative he had, lived in Kabul. He had not seen any of them in thirty years. They had limited contact via a phone call, once in a blue moon. He dreamed about one day seeing them. It was a shame he was only three hours away but couldn't tell them.

"Why can't you tell them?"

"Our company told us we could get in trouble if we contacted our family in Afghanistan."

"What kind of policy is that? How can they tell you not to talk to your own family? Besides, why don't you just go and visit them on your first vacation?"

"We can't do that either. It's also against policy for us to take vacation in Afghanistan," Mr. G sadly replied.

Americans and our screwed up ways, I thought. *Operational security* is always the excuse for everything. Remember, fear makes money. If people don't remain in fear, nobody makes the big money. If interpreters start visiting family all over the country, it would appear as if no one is scared anymore. But, how can a private company dictate where its employees travel on their vacation period? Is that even legal? Even though they are originally from Afghanistan, most interpreters like Mr. G, are U.S. citizens with the same rights as you and me. No private company is going to dictate where this U.S. citizen travels, I can assure you. As long as it's within United States law, it's none of their damn business.

"Mr. G, that's ridiculous. I'm going to make you this promise. I can't tell you when it will happen or for how long. It might be next week, it might be next month, or it might be in six months. But when we get the time, I will take you to see your family in Kabul. Mission Essential can kiss my ass. It would be my honor to ensure that you get to see your family after thirty years."

Mr. G was in disbelief, but by then he knew how I operated, which certainly wasn't in regard to any policy.

"Can I call them?"

"Call them right now. Tell them where you are. Just don't talk about the team. Everything else is fine. Make sure you don't get their hopes up that we're coming anytime soon. Let them know that when we do come, it will be on short notice."

Mr. G talked to his family every day after that conversation. They were very excited to hear from him and the prospect of actually getting to see him. While I had to convince him that his visit would be a luxury and would take a backseat to our operations, I had secretly made it a priority. I needed to go to Kabul on legitimate business, anyway. We would handle our business and then stay the night at his family's home.

I briefed Sammy on the plan, but didn't tell Mr. G until the day before we were going to leave. I wanted to surprise him.

"Mr. G, pack your bags. At 6:00 a.m. tomorrow, we're going to Kabul. Tell your family we'll be staying the night with them after we conduct our business. We'll have to get up and leave the next morning, so it will be a short visit. And tell them we're coming hungry."

Mr. G was excited and started in with me about going shopping beforehand, so he could buy some good clothes. His metro-sexual ways were ingrained. I explained that we didn't have time to go shopping. His family would welcome him all the same, in the man-jams he had.

The next day, we rolled into Kabul and handled some administrative issues. By 4:00 p.m., we pulled in front of the home belonging to Mr. G's aunt. We were met by two small children who immediately gave him big hugs. The kids had never seen him before but had been told he was family. That's all they needed.

As we entered the courtyard, the reunion began. It was enough to bring tears to the eyes of even the most hardened person. Sammy and I took pictures of Mr. G and his aunt embracing, after thirty years on opposite sides of the globe. I felt a sense of accomplishment that would go down as one of the highlights of my tour. We were treated like kings the entire time we were there.

I would come to love Mr. G like a brother on a personal level. On a professional level, he would piss me off to no end. He just wasn't cut out for the position. In his defense, not many people were. He was an ordinary civilian in the U.S. on one day, and the next, he had been thrown in the middle of an undercover counter-terrorism team in Afghanistan. It's hard to imagine how scared he must have been.

Mr. G would go on to win the FOB Fenty pool tournament, win every poker game he played in, and beat even the smartest ORSA (Operations Research and Systems Analysis) analyst every time at chess. Everyone loved being around him.

PIMP DOWN! Amir Jit Singh is arrested by ISU agents while carrying a remote control, magnetic IED, in a rice bag. Photo at top shows the "sticky bomb." Note the circular speaker magnets utilized for quick attachment to a vehicle. The remote control was found in his pocket.

With too much ordnance to carry back to the vehicles, a local villager and his donkey were commissioned to ferry charges, in order to blow the cache in place. Sammy and I took a moment to take a trophy picture with a portion of the haul.

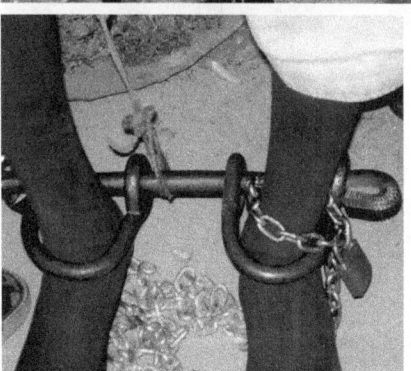

Hostage rescue operation is a success, but no one thought to bring the bolt cutters. The victim had to be carried down the mountain on the backs of the agents.

124 Killing Sheep: The Righteous Insurgent

Looking like this, the challenge is keeping the Americans from shooting at you as well. Below: Having a little bit of fun, playing the burned-out-gas-can cello, in the midst of a get-together.

Mr. Wakil the bomb maker (white man-jams), is forced to eat breakfast with his captors, which includes one gringo *crusader*. His perplexed look is because it's hard to hate a guy who just passed you the eggs and filled your tea glass. Below is his bomb making kit that was seized from his residence, by ISU agents.

M-9 Pistol?	Five hundred U.S. dollars
M-4 Rifle?	One thousand U.S. dollars
3,000 pounds of Hash?	Three million U.S. dollars

Stealing 3,000 pounds of hash from Taliban forces during a midnight ambush, turning their mules loose, and then using the Taliban's hash as cover for two hard-hitting crusaders, all the while being less than ten miles from the Pakistan border and Tora Bora?

PRICELESS!

Operation Night Train is a success.

Teaching my Afghan brothers how to party—*redneck style*. The red plastic cups were a must. Below: A remnant of the Soviet invasion. In ten years, we will be taking pictures next to American MRAPs.

Counting beans and hating life. The only thing that could make the dog and pony shows of Campus Eggers half bearable, was the excitement of running beer and liquor at night. It was a tribute to my grandfather, who was a ridge runner and hauled moonshine back in the day. Good times!

CHAPTER 6

Cowboys and Indians

In January, Agent S received information from one of his informants, about a cell involved in manufacturing and distributing bombs (IEDs). The informant told us a man in Chaparhar District was building the bombs and passing them off to an Indian guy in Jalalabad. The Indian guy was selling them to whoever had cash money.

The story sounded a little bit far-fetched, initially. But in Jalalabad, there is a substantial Hindu and Sikh community. They live in harmony among the Afghans and even have a temple they attend. To my knowledge, the temple has never been the subject of any attacks. For some reason, they just get along. But, the idea of an Indian working in conjunction with a Muslim to distribute bombs just didn't sound right.

The informant would keep in contact with Agent S and call him as soon as he could find out more information. The problem was, the informant was not actually a part of the organization. He was relying on limited information he was getting from an acquaintance directly familiar with the cell.

A few days later, the informant called back with details on the Indian gentleman's residence in Jalalabad. He also heard that a bomb had been delivered to the guy earlier in the day. The bomb was reportedly going to be hidden in the temple. He gave a description of a Toyota Corolla the Indian drove. That's all he knew, but it was vital information. It gave us a starting point with which to work our magic.

There was no time like the present, so we went to work. We formulated a hasty plan of action in order to locate the Indian and figure out what he was up to. We would establish surveillance on the temple itself, the Indian's residence, and a traffic circle that was halfway between the two locations. Since we had no other specific information, we would let the scene play out and make decisions based off our observations. Although the information indicated the bomb was going to be hidden in the temple, we couldn't go inside. It wasn't within the culture. Absent any solid emergency, Muslims would not go inside the temple out of respect. The same way Americans can't go into a mosque—it's just not done.

"Sir, the colonel will go talk to the general and get some CID agents to go with us on surveillance," Sammy advised.

"Sounds good, my friend. Let's get radios issued out and get everyone on the street."

With that, I made certain everyone involved had a portable radio, their pistols, and at least one rifle in each vehicle. We would be in the city with plenty of backup if we needed it. On this operation there was no need for everyone to carry a rifle. Half of the troops would be deployed on foot, anyway. A few minutes later, Forerunners rolled out of the safe house on the hunt for a merchant of death.

Agents took up their positions around the locations, some in vehicles, and some on foot. Sammy and I kept watch from inside our Forerunner. Radio contact was established with everyone involved, at which point the stage was set. A Toyota Corolla, matching the description given by the informant, was parked outside the temple. It could possibly belong to our target. All we could do now was sit back and wait. In the States, there were days and even weeks, where I basically lived out of my car while sitting on surveillance. Surveillance can be hours upon hours of boredom, with a few minutes of action. Or, things can happen as soon as you get set up. It's a crapshoot.

After a couple of hours, we observed an Indian male exit the temple, and get into the Toyota Corolla we were watching.

"I've got a potential bad guy getting into the Corolla. He's an Indian male, wearing a red turban, white clothes, and a black jacket. He's not carrying anything," crackled the radio.

"That's probably our guy. I'll go with him," was the response.

The man drove away, and passed by the agents on surveillance at the traffic circle. "OK, he just passed us, heading in the direction of the target house."

A few minutes later, the radio crackled again, and an agent told everyone that the Toyota was pulling into the target residence. It appeared as if we already had our man in our sights.

"He parked the car and is walking inside. Not carrying anything," the next transmission advised.

Thirty minutes later, "He's leaving the house in a different car. It's a blue Corolla. I couldn't tell if he was carrying anything. We're all going to go with him."

The Indian once again passed by the traffic circle, but this time he unknowingly had several agents in tow. He drove straight back to the temple and parked down the street. The entire team was on him at that point in the game.

"OK, he's getting out…looking around…reaching back into the car…now he's back to looking around. This bastard is nervous as shit… OK, he's reaching back into the car…OK, he's got a rice bag and he's being very careful with it. This looks good!"

"He shut the door…now he's walking toward the temple!"

"Let's take him!"

The situation looked good. The information had jived so far with everything the informant had told us. We decided to roll the dice and take him down. This was Afghanistan, not the States. We could err on the side of caution and get away with it. No one was going to scream racial discrimination or profiling. *You only stopped me because I'm Indian*, would not be an issue.

Agent N was only a few steps behind the Indian soon after he left the Corolla. Agent N waited until he was away from a crowd of people and then struck. He pulled out the small Makarov pistol concealed under his man-jams and stuck it to the Indian's head.

"Police! Don't move! If you move, I will kill you! Slowly, drop that *fucking* bag!"

The Indian had two choices. The first was to comply. The second was to die. He wisely chose to comply and slowly sat the rice bag down on the pavement. The way he sat it down told us there probably was a bomb in the bag. As soon as the bag was on the pavement, Agent N took the guy down hard. There was no use in taking any chances on this type of case. The poor Indian didn't have time to do anything but see stars, get acquainted with the concrete, and try to maintain consciousness. He was immediately surrounded by the rest of the team and handcuffed.

A quick look into the rice bag confirmed that it contained a bomb. That was all I needed to see. The EOD people could do any further looking as far as I was concerned. We were agents, not bomb techs. I knew my limitations and had no desire to die by way of explosion.

Sammy and I grabbed the Indian and walked him over to our vehicle. A search of his person revealed a remote control doorbell actuator, which would have been used to trigger the bomb. Luckily, the two AAA batteries were not installed in the remote. The batteries were loose in his pocket so the bomb would not inadvertently detonate prior to deployment. It wasn't exactly an effective safety mechanism, considering that all of those particular remote controls were usually on the same frequency. Had someone rang a doorbell in the near vicinity, we all would have been riding the paradise express.

Due to all of the excitement, people started to gather in the immediate area. Uniformed officers arrived on the scene, and were doing their best with crowd control and the diversion of traffic. However, Afghan citizens aren't accustomed to crime scene tape. They're curious and want to see what's going on. If you don't believe me, just get into a car accident in Afghanistan. Within five minutes, there will be thirty people there arguing about who was at fault, even though most of them didn't even witness the crash.

"Sammy, tell them to keep the area blocked off and for no one to mess with it. This thing is obviously radio controlled so let's call the Americans. They can handle this one."

Sammy relayed the instructions. Everyone knew it would take the Americans a lifetime before they finally arrived. It was going be a long night for all involved.

"Let's get this guy off the scene and over to counter-terrorism. They can interrogate him while we're dealing with the bomb."

With that, we loaded up the Indian in our vehicle, and along with Agent A, transported him to the Counter-Terrorism Division. It only took a few minutes to get there. As they escorted the Indian into the building, I began making my calls to Task Force Paladin. I asked them to contact Task Force Gryphon and request assistance. All I knew was that we had a remote control bomb in a bag, sitting on a street corner in the middle of town. They needed to hurry because it was a difficult situation trying to divert traffic and control the crowd.

Through a series of phone calls, unnecessary questions, and haggling, Paladin finally advised that Task Force Gryphon was dispatching an EOD unit. I was able to talk to the young EOD buck sergeant (via cell phone) who would be in charge of the scene.

"My friend, all I can tell you is that it's a bomb in a rice bag. We recovered a doorbell remote from the bad guy's pocket. After a quick peek to confirm, no one has messed with it since. It's on a street corner in the middle of town and the ANP are doing their best to hold a cordon."

I wished I had more information to give him, but it was a textbook operation. We had arrested the bad guy, no one had screwed with the bomb, and we had set a cordon. There wasn't anything else we could or should do. Any further action would rest squarely on his shoulders. I told him to be safe and that I would hook up with him later to compare notes.

● ● ●

I sat in on the interview of the gentleman, whose name was Amir Jit Singh. As with a lot of bad guys you deal with, Mr. Singh had the utmost inability to tell the truth. The only thing he was truthful about was that there was indeed, a bomb in the bag. His story started out that he was going to kill some people who were involved with the death of his father and his brother. He then claimed he was going to kill someone involved with the arrest of his brother in India, over a heroin transaction. He said he paid thirty thousand Pakistani rupees for the bomb. He confirmed that if the two AAA batteries were inserted in the remote, a single push of the doorbell button would trigger the bomb and cause it to detonate.

Mr. Singh was full of stories. What he was doing, was trying to minimize his involvement with the IED cell. He wanted us to think that he had bought the bomb to settle a personal dispute, which was an isolated incident. Criminals often admit to lesser crimes, in order to avert attention away from their more sinister activity. It wasn't working. We hadn't gotten on to him because of a personal dispute. Our attention had been brought to him because he was reportedly selling bombs in Jalalabad. Mr. Singh just didn't want to give up the rest of the cell. While it wasn't helping our case, it was understandable because he had a family to worry about. His accomplices might kill his family if he snitched. That was a true possibility. We weren't in the States.

While the interview was getting us nowhere in further identifying his crew, the gringos had finally arrived on the scene. While it took them over an hour to go one mile away from their prison camp, it only took a short time for the skilled EOD tech to make quick work of our bomb. He disarmed it and allowed our guys to take some pictures. After collecting the components, he and his team packed up and headed to the house. It was just another call for the young crew.

I told Sammy that since the interview was going nowhere, we might as well cut everyone loose. It was a good day's work for everyone involved. We had one bomb distributor in custody, caught red-handed, and one bomb off the streets. Whoever was making them would now have to find a new retailer to work with. Hopefully, some fingerprints would come back off the components collected by EOD, which would further our investigation. Plus, we still had the snitch in place. We weren't quite finished with the cell just yet.

● ● ●

Sammy and I left Mr. Singh in the capable hands of the counter-terrorism colonel and headed toward FOB Finley-Shields to link up with the EOD crew. I had collected the doorbell remote and the batteries taken from Mr. Singh, to give to EOD for submission to the lab at Bagram. I also needed to get some more information for my report.

FOB Finley-Shields and Camp Hughie are located adjacent to one another, off the main road that leads into Jalalabad. They are only about a half-mile from FOB Fenty. Technically, they are located on Afghan National Army property, therefore, the ANA soldiers provide exterior security. They obviously are much more militant than dealing with the police or private security guards.

As we approached the gate, I turned on the dome light in the Forerunner so the guards could see us holding up our identification cards. When approaching a checkpoint at night, it is customary for the driver to turn off the vehicle's headlights. This is commonsense and allows the guards to see inside the vehicle while not blinding them. The problem with our vehicle was that the lights were jacked up. You couldn't turn off the headlights when the engine was running. With all of our other vehicles, you could. While we often transited ANA bases, it was never fun at night, especially in our ride.

"Here we go, let's see what happens this time," I said to Sammy.

As I crept toward the entry control point, with the intent of stopping well ahead and killing the engine, I had already gotten too close.

"*Drayyyysh!!!!!*" the young soldiers screamed in unison, "*Drayyyysh!!!!!*"

Two young ANA soldiers were on us. One AK-47 was pointed at me and another was aimed at Sammy. This scenario happened to us on a routine basis because we looked like hillbillies, but it was never a routine feeling when it came to staring down the barrel of a rifle at night.

Sammy calmly went through the routine of explaining who we were and where we needed to go. It was a rare occasion that soldiers or police officers did not know Sammy. This was one of those rare occasions. After inspecting our identification cards, the two young soldiers relaxed. They called for their commander to come to the gate and clear us.

A few minutes later, the commander arrived. This time it was a familiar face. Sammy and I exchanged greetings with the ANA commander and conversed for several minutes. The ANA commander told us to just follow him so that no one would give us any problems at the next checkpoint. He escorted us to the building we needed to get to, and then went about his way. I told Sammy to wait in the vehicle and that I would be right back. I called the young EOD sergeant on my cell phone and he indicated he would be right out to meet me. A few minutes later, he emerged from the operations center.

I think there was some initial shock as we shook hands, because standing before him was a gentleman who appeared to be Afghan. I hadn't gone through the routine of changing clothes because I was too tired to make the effort. Besides, there was no mistaking my southern U.S. accent. Appearances can be deceiving, but no Afghan can mimic the way I talk. After a brief conversation and a few laughs about my attire, we got down to business.

"You really pissed off the colonel. She's probably going to talk to the brigade commander about what happened tonight," were his first words relating to the subject at hand.

"Who's pissed off?" I asked, even though I already knew who the culprit was. I played dumb to see what he would say.

"Lieutenant Colonel N, the battlespace owner. She's pissed because you guys didn't clear the operation with her first."

"Let me clear something up real quick, my friend. I don't work for her and neither do my people. I don't give a rat's ass about what she thinks," I said.

The young sergeant was surprised at my reaction to his statement. It wasn't the typical reaction one would display, after finding out that a battlespace renter was upset at you. Anyone else would have stood there shaking in their boots because a lieutenant colonel knew their name, but I was too tired to care about hurt egos.

...The reality of this situation was that the Afghan police had conducted an investigation. It required neither the approval, nor the consent of the United States military. It was, and is, their country. I felt both anger and disappointment, as I pondered the first reaction to us grabbing a high-level member of an IED cell. While I wasn't expecting a medal, the appropriate course of action should have been for her to congratulate the Afghans on a job well done.

"Don't worry about the politics of the situation because it's above your head. Let us old folks bicker about that. You gents merely did your job so don't sweat it. We appreciate you guys coming out and risking your lives, screwing with that bomb." I tried to downplay the politics of the situation, but inside my head, I knew a shitstorm was awaiting my arrival back at FOB Fenty.

"Tell me about the bomb," I said, as I handed him the bag with the remote control in it.

"It was a magnetic, radio controlled IED. It had three speaker magnets attached to it. It's the same kind they've attached to the backs of fuel trucks about to enter FOB Fenty."

"So, it was a sticky bomb?"

"Yeah," he chuckled.

Military folks don't use the term "sticky bomb." They don't even say the word "bomb." He was amused at my simplicity and layman's terms.

"How did you find the guy who had it?"

"It's called police work, my friend. I'll tell you all about it later but right now I've got to roll."

I made a few notes for my report, shook hands with the young EOD sergeant, and humped back to the vehicle.

"Would you believe this crap? The bitch is pissed because we didn't clear the operation with her first. Now I'm going to have to deal with a load of crap back on the FOB. You'll have to hold things down tomorrow because I'm sure I'll be busy kissing ass all day."

"Why are they unhappy? We just captured a bomber!" Sammy exclaimed, not having a true understanding of American politics.

"Somebody's feelings are hurt because they think they control everything that happens in their battlespace. We both know they don't control one thing outside the FOB, but you can't tell them that."

Sammy, as with anyone who does a good job, thought that words of praise would accompany this case. It bewildered him as to why the Americans wouldn't be happy about what we had accomplished.

"Just forget it. As a buddy in Iraq once told me, don't try to change the river. Let's go get a beer."

"No problem," was Sammy's only reply, as he cracked a smile.

As we headed over to our favorite shop in the bazaar, I spent the time trying to explain to Sammy what I meant when I said, *Don't try to change the river.* When we finally pulled up in front of the store, he said that he understood the concept, but I think he was just in a hurry to get to drinking. We sat there in the Forerunner, sipping on a couple of warm Heinekens while watching the traffic pass by. If I didn't have to return to the prison camp to face the scrutiny of the gringos, it would have been a perfect day. I dropped Sammy off at his house and made my way back through town. With only about twenty minutes of being hassled at the checkpoint, I was finally back at Paladin and in my cubicle. The next morning, the drama began. While I wasn't privy to all the e-mail traffic, phone calls, and meetings that might have taken place, it was obvious LTC N and her crew had stirred the pot. In fact, after looking at the times on the e-mails, they were stirring the pot while me and my crew were still out on the case. (E-mails cited within this book are quoted as they originally appeared, with some minor redactions and changes to format. The redactions are irrelevant to the message of the text. Names have been redacted with asterisks, in order to protect the guilty.)

> Gentlemen,
>
> I know you have mentioned based on my previous messages that a policy regarding what the real policy is behind **** and their "collection" and so on. However, we've had another issue that is once again highlighting why a comprehensive policy needs to be made in writing on what **** can or cannot do needs to be made and as soon as possible. In this case, our **** working with the ISP program, mr. Blanchard, was apparently out and about with the Afghans completely on his own in essentially an "undercover" role without any other Coalition Forces present at all. He supposedly even almost got detained by the ANA. This cannot keep going on without something that shows where the lines are in terms of their ability to "collect" intelligence information, criminal interviews/interrogations...
>
> Respectfully,
> CPT G***

Let me evaluate this initial e-mail before I continue. Take notice of how Captain G utilized quotations around the words *collection, undercover,* and *collect*. This is a subtle method to show a lack of respect and a lack of acknowledgement for our operation.

First of all, what were we *collecting*? That's an intelligence term, not a law enforcement term. Intelligence people collect information and write reports. We're cops. Information to us is useless unless you put it to work. The only thing we *collected* was a bomb and a bad guy. We *deposited* said bad guy in the jailhouse, while EOD destroyed the bomb.

Those were key facts Captain G chose to leave out of his e-mail. Had he included that slightly pertinent information, maybe somebody on up the chain would have said, *Wait a minute. Stop crying. These guys took out an IED cell, so why are we bickering about this? Don't you have better things to do?"*

Where did the story come from that I was almost detained by the Afghan army? Somebody had to make that up. That's comical, considering I drink beer with several of the local ANA commanders. The only way I'd get detained by the ANA, is if I showed up to a get-together without that bottle of Jack Daniel's I'd been promising one of the colonels. That's ground truth information Captain G wouldn't know while sitting behind his computer on the prison camp.

...In Captain G's defense, I'm going to view the situation from his perspective, and try to explain why he was unhappy. I'm going to even offer this gentleman some advice. First of all, Captain G is the S-2X, which means he's an intelligence officer who coordinates human intelligence (HUMINT) and counter-intelligence. He is reportedly the focal point for source operations (running informants). Now, military culture in general, dictates that if anything happens within your lane, you should know about it. If you find out about something after the fact, it's taken personal. This creates a persona of arrogance, because you begin to feel that you are entitled to the information, and that it's not your job to seek it out. I'm sure that since Captain G was the S-2X, he felt that everyone should come to him, instead of him having to solicit the information from others. That mentality might suffice, if you are only concerned with folks within your chain of command. You can hold them accountable. Captain G's arrogance prevented him from considering that there might be other entities, within his area of operations, that were conducting similar business—entities that had no mandate to report to him. Entities that had never heard of him. Entities that didn't even knew what S-2X meant. Therefore, he had two courses of action.

First, he could have made an attempt to develop relationships with the locals involved in police work and intelligence collection. This would have opened lines of communication. He probably would have been kept in the loop on their operations (ours included), if he had done so. This would have required a bit of effort, but making friends isn't that difficult. Just drop the elitist attitude and get to know the locals.

Second, he could have chosen to do nothing, and just expect everyone to acknowledge his position. In our case, that was the course of action he took. The problem for him was that the Afghans were not within his chain of command. The Afghan police do not answer to the U.S. military. Therefore, the effects of his attitude resulted in him being the last one to hear about things. We used to have a saying in narcotics work, whenever a turf battle developed over a case, "You can either join us on the case, or read about it in the newspaper." Captain G found himself reading about this one in the newspaper. Some nasty old Afghans and a nasty old contractor had stepped into his lane, without his permission. It made him look bad and hurt his ego. For that, I do feel sorry for him, but it wasn't our intentions, nor was it our concern. We had a job to do. If Captain G wanted to get on board and join us, we would have welcomed him as a brother. Relationships outweigh idle threats via e-mail—every time.

Since an e-mail had been generated, everyone who was copied would now take his remarks as fact. After all, he was the Brigade S2X. The second e-mail in the string was non-eventful.

> Gentlemen, once again the collection activities of a **** is being called into question. It would help us out if we could implement some guidelines for the BCTs sooner.
>
> Thanks for the assist.
> R,
>
>
> LTC ******

I don't have any issue with the brief comments of the LTC. While I don't know this gentleman, it was apparent he had more important things to do. He sent the e-mail to Gene ****** who was a Company X employee, serving as the country manager. But I wonder if he would have even forwarded this e-mail, if Captain G had mentioned the reason for me being out and about with those scary Afghans.

Gene's response to the e-mail was politically correct, but with a tone that said, *Hold on a minute, it sounds like you guys don't have your story straight.* While the e-mail is what you should expect from someone in his position, I felt like he stood up for me. Gene was a good man, and everyone liked him.

Thank you for bringing this to my attention. As you know, **** Mark Blackard is assigned to the CIED-IS Program at FOB Salerno under TF Paladin. There is in fact written policy and guidelines with regard to CIED-IS and all other **** activity. The CIED-IS Program is sometimes misunderstood. It is important to have all the facts with regard to any particular operation. Please allow me some time to gather those facts regarding this incident and circle back.

Thank-you,

Gene ******
Country Manager (GS15 EQ)
**** Program

Nowhere in the entire e-mail string was a mention of what we had actually done. There was only bickering over hurt feelings. In the midst of this storm of derogatory e-mails, I sent out my report on the operation to everyone on my distribution list. The response was what I had hoped for. People congratulated my team on a job well done. They wanted to know more about Mr. Singh, and why I thought an Indian had partnered with Muslims to sell bombs. That wasn't something they'd seen before.

Apparently, several higher-ranking officers had read the report and began to pump it out as a success story. That's exactly what I needed; some impromptu top-cover.

I wanted to stay under the radar. I elected to not cross the paths of the brigade and battalion level complainers, with the congratulatory folks at division. If it affected my position, I would play that card, but not unless I had to. How can you punish a person, when your bosses are reporting the operation as a good example of how a program is working? All it would have taken was for me to blast the negative e-mail string up to division. Most people probably would have. Besides, the unnecessary drama wouldn't have effected my fitness report. I had to lay low for a few days to let the smoke clear. Once I was sure I was no longer the topic for discussion, it was time to get back to work.

CHAPTER 7

Wake Up, Mr. Bomb Maker

On March 4th, the EOD first sergeant caught me as I walked into the tactical operations center (TOC).

"Hey Mark, see if you and your Dirty Dozen can find this turd. He's been building IEDs and hurting my soldiers."

The first sergeant was sitting at his computer in the row of desks behind my work station. I turned around and he handed me a four-page document that turned out to be a scrubbed down, unclassified prosecution packet. The first sergeant had come across it in his e-mail box. It had been sent out several weeks prior.

On January 15th, EOD personnel had responded to a report of a bomb in the roadway on a major supply route (MSR) in Chaparhar District. The EOD techs made quick work of the bomb, which was made out of a 107mm Chinese rocket. The rocket had a remote control device attached to it that served as the trigger. The EOD techs collected the components and sent them to the Combined Explosives Exploitation Cell on Bagram, known as CEXC, which is pronounced as "sexy." CEXC personnel were able to recover a fingerprint from the tape holding the remote control device to the rocket. They matched the fingerprint to an Afghan named Wakil.

The prosecution packet had a very simple diagram which showed the IED that had been disarmed. It had a close up of the fingerprint and a picture of Wakil. Since the packet was meant for release to the Afghans, it didn't contain any further information. However, that was all the Dirty Dozen and I needed. We had Wakil's fingerprints on a bomb which had been planted on the side of the road. That fact gave us probable cause to go and arrest him. We had his photograph, so we knew exactly who we were looking for.

"This gentleman just became public enemy number one. His time is limited. He just doesn't know it yet."

My response was a bit cocky, but I was going to prove a point. If the military, the intelligence folks, or anyone else asked for our help, we would deliver.

"He's definitely responsible for this IED. But I think he's behind a bunch more, because the others were all built the same way and went off in the same area," said the first sergeant.

I spent the rest of the night trying to dig up everything I could on Wakil. The only information I was able to glean was that village elders had turned him over to coalition forces a couple of years before, because they suspected him of making bombs. His fingerprints and photograph had been enrolled in the BATS (Biometric Automated Toolset) system during this incident. The BATS is a system used by the military to catalog fingerprints, names, photos, and iris scans, in order to positively identify a person. It's very similar to our system in the States known as IAFIS (Integrated Automated Fingerprint Identification System). I couldn't figure out why I was unable to find any further information. If he had been sent to the American detention center on Bagram, there should be more. It just wasn't there. I made enough copies of the prosecution packet for every agent on the team. I couldn't wait to get to the safe house the next morning, and brief the men.

At 7:00 a.m., I was honking the horn for the young soldier to open the gate. Sammy and most of the crew were already there, getting ready to start the day. After drinking tea and making small talk, it was time for the 8:00 a.m. meeting. As Sammy and I passed out the packets on Wakil, I explained what I needed to happen.

"Gentlemen, whatever you're working on right now, I want you to stop. Put it on hold. I don't care what it is. I want everyone on the team focused on one mission. That mission is to put this guy in handcuffs."

As Sammy translated my request, the agents were busy thumbing through the packets. Sammy made sure everyone understood that Wakil was a dangerous guy. He wasn't making simplistic pressure plate bombs. He was making remote control bombs and was obviously proficient in electronics. He was definitely no fool.

"The only two clues I can give you are the location from where the bomb was found in the road, and the village near where he was detained over two years ago. Other than that, it's up to you to find him," I advised, while the team members took notes.

We broke out the map and everyone gathered around. We poured over the location where the bomb had been disarmed. It was in the vicinity of Dagu Village.

"Gentlemen, as soon as we locate this asshole, a barbecue is in order and the beer is on me," I said.

After the meeting was adjourned, Sammy and I spent the rest of the day going over administrative issues with the colonel. I finally had enough of looking at spreadsheets and decided to call it a day.

Back at the Paladin compound, I ran into the first sergeant.

"Hey top, old Wakil is now being hunted. I'll keep you posted."

"Cool, man. I hope you guys can get him."

"Only a matter of time."

● ● ●

Two days later, Sammy and I decided to eat some free food at the chow hall for lunch. As we were eating, Sammy received a call from Lieutenant Colonel N. It only rang once. In Afghanistan, incoming calls are free on Roshan cell phones. Therefore, what people will do is call and hang up, forcing you to call them back and foot the bill. Afghans and rednecks are just alike.

Sammy called the lieutenant colonel back. His eyes got wider and he broke a smile during the conversation.

"Agent D has a snitch into Wakil."

"No way. That's too good to be true. Nothing's ever that easy," I said, while devouring a greasy pork chop.

"He does. He's going to meet with the CI this afternoon to show him the picture and make sure it's the right guy. They'll call us as soon as Agent D gets done meeting with him."

I could barely contain my excitement. A case like this was the exact purpose for my being in Afghanistan, making $250,000 U.S. dollars a year, courtesy of the taxpayers.

I wanted to go tell everyone I knew about this development, but I had already run my mouth enough. I don't like to cry wolf. I wanted to have him in handcuffs in my Forerunner, before I revealed anymore of what we had going on. The more people who knew about what we were doing, the greater chance we had of being suppressed. Or worse, the gringos might actually try to get involved and screw things up.

At around 4:00 p.m., the call came in. The informant had positively identified Wakil and provided a plethora of background information on him. According to the informant, Wakil still lived in Dagu Village. Wakil also had a brother who was locked up in Jalalabad for making bombs. Everyone in the village knew what Wakil was up to, but were deathly afraid of him. Two days before, Wakil had lured a villager out into the fields. Wakil suspected the villager was an informant for the National Directorate of Security (NDS), which is the Afghan equivalent of the CIA. Wakil shot and killed the guy. No one in the village was brave enough to report it to the ANP. Three months prior to this incident, Wakil's nephew died in a car bomb, while trying to attack a German patrol in Kunduz Province. Wakil wasn't a nice guy. He was dangerous.

According to Agent D, the informant was scared to death. That was a good sign he was telling the truth. Sammy and I immediately rolled out and met up with Lieutenant Colonel N (Colonel G had to go to Kabul on administrative matters and was not in town), Agent D, and the snitch. The snitch agreed to show us exactly where Wakil lived, but would do no more. He had a family to think about, and would not get caught doing surveillance or reporting on Wakil's whereabouts. He was a simple man. I didn't blame him for not wanting to get too involved. We loaded him into the hatch of my Forerunner and covered him with a blanket. Upon driving to the general area, he would direct us from the safety and concealment of tinted windows and the blanket.

As I turned off the paved road, I noticed that a Corolla had just traveled the same dirt we were on. That gave me a sense of relief. No pressure plates to deal with. We should be able to pull off the brief trip without raising anyone's suspicions. As we drove down the winding dirt road toward Dagu Village, we saw nothing but open fields on both sides. Villagers went about their work, paying little attention to my silver chariot with our precious cargo in the back.

As we pierced the edge of the village, the informant rose up from his position just enough to see out the side window. Moments later, he identified the target location.

"Up on the right. The green door right there."

With that, the informant retreated back under his blanket. He had done his part. After turning around and heading back out, Sammy locked in the GPS coordinates as we passed the qalat. We dropped the informant off at a taxi stand on the outskirts of Jalalabad and thanked him for his service. We would get with him later.

We spent the next morning going over possible scenarios as to how we would put handcuffs on our target. While Wakil and a few of his cronies with AK-47s would be no match for the raid team we would assemble, the concern was for the potential of remote control bombs along the road leading into the village. This guy wasn't an amateur. He was a professional bomb maker who was proficient in electronics. It would be nothing for him to mine the road leading to his home, to mine the perimeter of his home, or to mine areas within the qalat. If the raid team was detected by Wakil or his crew while on approach, it would be nothing for them to hit the button on a remote control and kill a few of us. There needed to be an element of surprise on this one.

After quarterbacking several scenarios, it was decided that Agent D would immediately deploy undercover, into the village. Agent D would be able to get eyes on Wakil's residence and provide some ground truth as to what was going on in the village. The entire team would remain in a high alert status while Agent D worked his magic. In the States, we would always set up in the immediate area to provide backup for the undercover agent. But in Afghanistan, it's not always feasible. In most areas it's like trying to do surveillance in a trailer park where everyone knows everyone.

I issued Agent D thirty U.S. dollars, pulled from official funds from the Bank of Mark (my own money from my right pocket). He needed a few bucks to buy some fruit, vegetables, and bread. Agent D would take the food to the CI's house, as if he were just a visiting relative. He also needed money for the taxi to get there. They were all legitimate expenses we couldn't seem to obtain funding for, from anywhere else. Billions of dollars were being spent on useless programs and I couldn't get thirty bucks to help catch a bomb maker.

After conducting a safety brief and discussing all the "what if's" in case Agent D should run into trouble, Sammy and I drove him to the taxi area in town.

"Be safe, my friend. Keep your cell phone on. If anything should break bad, we're coming in strong."

"OK, OK. No problem."

Agent D wasn't the least bit concerned. To anyone in the village, he was just visiting. There was no need for me to be making a big deal out of it. Just give him some room to work.

"And don't be drinking on the way over there!"

We laughed. We all knew Agent D was going to have a Heineken or two for the road. After all, it was a good fifteen-minute ride in the taxi. That made for plenty of time to have a drink, by his playbook.

Sammy and I returned to the safe house to remain on standby with the rest of the team. The routine of equipment checks, weapons checks, and radio checks began. As I commenced my inspection of the vehicles, I was happy to find that the tanks were topped off and all of the safety equipment was present. The young soldier stood by proudly because he knew everything in his area of responsibility was squared away.

I began checking the individual agents and found that everyone was loaded for bear. Everyone had a radio, all of their magazines, and the rest of the gear I always insisted they carry. It was now a waiting game to see what Agent D had brewing in Dagu Village.

At around two in the afternoon, Agent D called and reported that he had eyes on Wakil's residence. Several individuals had come and gone but there was no sign of Wakil. He would call again, soon.

We sat out on the front porch killing time and enjoying general conversation. When trying to catch bad guys, there is always a lot of hurry up and wait to deal with. At about three-thirty, Agent D phoned the lieutenant colonel.

"He's got eyes on Wakil as we speak," the lieutenant colonel relayed.

At the conclusion of the conversation, the LTC briefed everyone. Agent D had positively identified Wakil, who had emerged from the qalat and was working in a nearby field. The field was in a bad location because Wakil would be able to see us coming a mile out. He could easily flee on foot and disappear into the woods. A snatch and grab wasn't going to work under those circumstances. However, if for some reason he ventured out to the main road or to any of the small shops in the area, a snatch and grab would be possible. Agent D would stick to Wakil and update us with his movements. The snatch and grab was designated the hasty plan of action, in case an opportunity presented itself in the interim. If Wakil walked toward the roadway, the plan would be triggered.

The primary plan we formulated was an early morning raid on Wakil's residence, the next day. There was no use in wasting any more time. We had the guy in our sights and had to act quickly. In Afghanistan, if a bad guy gets wind he's being hunted, he can disappear into Pakistan and carry on. We weren't going to let this particular idiot even make it out of the village. An operations briefing was conducted. Sammy drew out an excellent diagram by hand, of the target qalat and the route we would take to get there. We would stage the vehicles just off of the paved road, and patrol down into the village. It was too dangerous to try and drive because the noise would give up the element of surprise.

Lieutenant Colonel N, Sammy, and I rolled to the PHQ to meet with the general and round up a posse. The general was happy to see us as always. After good conversation on a personal level and a few rounds of tea, we got down to business.

"General, this guy is the real deal. He's a good bomb maker and has hurt a lot of people, mostly Afghans, but a few Americans as well. I've got to be honest. It might get ugly on this one. This guy's dangerous."

"Shahuddin, I will have the counter-terrorism colonel prepare a team to assist you gentlemen. Inshallah, he will not resist."

We met with the counter-terrorism director at his office on the grounds. We briefed him on the operation and showed him Wakil's picture. His eyes grew wide and he chuckled.

"We too, have been looking for this Wakil. Now we will have him."

We all agreed to meet there at the PHQ, at 4:00 a.m. the next morning. Colonel H would have the raid team assembled and ready to go. Lieutenant Colonel N would contact a local prosecutor and have him on standby. Once again, the stage was set to do some damage.

...Lieutenant Colonel N was a strong leader and an operations type of guy. In his book, mission accomplishment always outweighed the welfare of the men. He never missed an operation the whole time I worked with him. He had worked for the Soviets during their incursion into Afghanistan, and spoke fluent Russian. He knew more about Soviet weapons, explosives, and ordnance, than any Russian ever thought about. He was the equivalent of an EOD technician, and was an invaluable member of the team. He never once displayed any sign of fear and was very well respected by the men.

● ● ●

Lieutenant Colonel N ordered the men to remain in place at the safe house, in lieu of going home for the night. I wanted to stay as well, but needed to verify there were no fires to be put out at the prison camp. Luckily, no one was complaining about me that night and the coast was clear to do what I was paid to do. I fueled up my vehicle and stole the typical complement of bottled water. After a couple of hours of sleep, I was headed back out the gate.

HONK! HONK! HONK! I blew the horn so the soldier would come and open the gate for me. When I first arrived in Afghanistan, I didn't like to blow the horn at night out of courtesy for the neighbors. I quickly learned it's not looked upon as being discourteous. Just blow the horn.

That's what it's for. The team was waking up when I pulled in. Soon, we were geared up and pulling into police headquarters. In the darkness of four in the morning, a small army had been assembled. There were young ANP officers standing in formation in front of the older convoy commander, who was barking out operational details. Other young officers were loading ammunition and RPG rounds into the back of the Ford Rangers. Machine gunners were inspecting their weapons mounted in the backs of the trucks. Police radios crackled in the night air.

Lieutenant Colonel N, Sammy, and I met with the convoy commander and Colonel H. Last minute details were worked out and contingency plans were discussed. The dispatchers were briefed and would be ready to send backup units, if necessary. The local district chief was standing by to assist and would send additional units upon our arrival.

At around 4:15 a.m., a convoy of twelve vehicles left the PHQ, again like the start of a NASCAR race. Our vehicles were in the lead since we had already conducted a recon of the area. The trip didn't take long, as we were the only ones on the roads. As we got close to the staging area, all vehicles killed their headlights except for the lead. In the darkness, it appeared as if a lone vehicle was traveling down the road. We pulled off of the pavement, only far enough down the dirt road to conceal the convoy from passing vehicles. Young ANP officers bailed out of the back of the pickup trucks and secured weapons. It seemed that one out of every four officers either had an RPG launcher or a PKM machine gun. As they assembled, I looked for Agent D. It made sense for him to lead the patrol.

He was nowhere to be found near the front of the action. Sammy inquired on the radio, and found out he was going to remain at the vehicle staging area.

"Why the hell is the case agent staying at the staging area?" I whispered to Sammy, in bewilderment.

"I don't want to say, but they are scared, sir. That's why Colonel G hasn't come back from Kabul yet."

"We're all scared, but this is our deal. There's no way those young kids are going down that dirt road in the lead. We started this shit."

With that, Sammy and I stepped out in front of the young ANP officers. I took the right side of the road and Sammy took the left. With a wave of my hand and a couple of taps on shoulders, a platoon-sized element began to follow in the darkness. Nearly a kilometer of ground had to be covered to reach the target.

"Everybody wants to go to heaven..."

"But nobody wants to go now..."

"*Fuck* it, I'm ready."

"I'm ready, too. Let's go."

We began to push down the road as quickly as we could without breaking into a run. We didn't have night vision goggles, so no one was peering through green to cut open the darkness. The wall of men, machine guns, RPG launchers, and ammunition moved silently. It was almost as if the young ANP officers thought the slightest sound would trigger an explosion. The uncertainty of the potential for mines scared the hell out of everyone. Nothing else seemed to matter except for the ground beneath our feet.

At 4:45 a.m., there was nothing moving but us. I stopped briefly at the halfway point to allow the slinky effect of a Ranger file to tighten up. During this pause, Lieutenant Colonel N joined us on point. There was no way he was going to be left out of any action.

The ground became the lesser threat as we approached the qalat. It was a big compound with walls over twenty feet in height. It was built like a small fortress.

We made it without encountering any resistance. So far, we were on track. We began to position officers around the qalat and into defensive positions. The opposition could come from inside the qalat or from the village itself. It took several minutes of hasty coordination via whispers and hand signals, but by 5:00 a.m., an effective perimeter had been established around the compound. The element of surprise was all ours.

Now, there is always more than one way to skin a cat. When you are new and inexperienced in law enforcement, or the military for that matter, you elect to do things heavy-handed. Kicking in doors and throwing people to the ground is a lot more fun than doing things low-key. It's sexier and it has the potential to generate medals. One gains respect among peers because he has been tested in combat. Therefore, *going dynamic* (kicking in the door) is usually the preferred method chosen by inexperienced decision makers struggling to define their commands. Everyone in the way would be considered collateral damage. That's not the way we looked at it, because everyone in charge had seen enough death to last them a lifetime.

We could have elected to hit the place strong and would have been righteous to do so. We had enough people to take it down, but we knew nothing about the inside of the qalat. Sure, I could have asked for some aerial imagery of the place, but pictures do not tell you how many people are inside, or where. We didn't know anything about how many women or children were present. We would have bust through the door blind. While it would have made for a great war-story and probably instigated a firefight, it wasn't the smart way to handle the situation. It could have ended up with a child or a woman being killed. They are merely innocents in these situations, and should not be blamed for the actions of the men in the household. Therefore, we owed them as much protection as any other citizen.

Once the perimeter was established, Lieutenant Colonel N contacted the prosecutor via cell phone and informed him of the situation. The prosecutor advised he would be on his way to the scene. This would ensure Wakil would be prosecuted once we grabbed him. There was no better way than for the prosecutor to see things first hand. If we had to get nasty, having the prosecutor approve of our actions would cover us with the courts, the governor, and the politicians in Kabul.

Sammy and I were covering the front door with two ANP officers. A young Afghan approached the perimeter to our left, and was told to halt. The man identified himself as Wakil's cousin and confirmed that Wakil was indeed, inside the compound.

We had a rat in a box. There would be no escape.

The young Afghan departed and returned a few minutes later with one of the village elders. He looked like a wise man from a nativity scene. The village elder spoke to Lieutenant Colonel N and asked what was going on. LTC N told the old gentleman we were there for Wakil and that we weren't leaving without him. The elder asked if LTC N would allow him to help resolve the situation to avoid any conflict or gunfire. LTC N gave him the go-ahead.

With Sammy and me covering the old man, he strutted up and pounded on the metal door with his walking stick. The sound of solid wood hitting metal pierced the quietness of the early morning.

BANG! BANG! BANG! A few minutes later, the door opened and a young girl slightly emerged. The elder issued a direct order to the child. You didn't have to speak Pashtu to figure out what he said.

"Tell Wakil to get his ass out here immediately."

He probably didn't use the word *ass*, but that's how I interpreted it. Regardless of the exact translation, it was all he had to say. The little girl disappeared in a flash. The elder waited in front of the open door, not concerned with any potential threat to his safety. This was his village and he would not allow Wakil to cause any further bloodshed by resisting us.

Besides, we had empowered the old man. With us behind him, he was no longer afraid to stand up and get gangster with Wakil.

Three minutes later, our prey walked through the open door. He seemed as if he had been sleeping only moments before. Since we didn't have pre-raid surveillance on the location, it was entirely possible that Wakil had been out earlier in the night, getting into mischief. Sammy grabbed him while I covered the doorway with my rifle. The young Afghans emulated what I was doing, which allowed me to help Sammy get some flex cuffs on Wakil. After a quick search of Wakil's person, Sammy and I wasted no time getting him up to the vehicle staging area. The ANP officers would hold the perimeter until the prosecutor arrived. Obviously, there would be additional evidence in his house that needed to be seized. The elder would remain on the scene to issue orders and keep the peace.

The walk back up the dirt road was a lot less stressful and frightening, when compared to the walk down in the darkness. Due to the number of shooters we had assembled, anyone would have been stupid to take a pop shot at us. Several villagers gazed upon us, as we did what amounted to a perp-walk, back up to the vehicles. We purposely made it a spectacle because there was a bigger picture to this operation. An element that really only Sammy and I understood.

While the immediate objective was to catch a bomb maker, the more important mission was to instill confidence in the community that the police and the government were in charge—not the bad guys. That's the true definition and objective of counter-insurgency. We wanted the people to see that the police were following the law and were not corrupt. More importantly, that the police were capable of doing their job and protecting the public without the U.S. military being involved. The people needed to know that the police could go wherever the hell they wanted to, when they wanted to. It would also send a warning to other crooks in the village that they could be snatched up as well.

As we passed officers along the route, they all perked up when they realized we had our man. Everyone felt a sense of accomplishment. This asshole wouldn't be building any more bombs that could hurt innocent civilians in Nangarhar Province.

When we made it to the vehicle staging area, Wakil was handed off to uniformed personnel for a photo shoot. I wanted to make sure it was publicized among the gringos that this was an Afghan operation. They were handling the dirty work and planning their own missions. At the time, this contradicted what the media and the intelligence community were reporting.

Two prosecutors arrived on scene and conferred with Lieutenant Colonel N. They were given the scrubbed down prosecution package that had been prepared by the folks at Bagram. After some clarification, the prosecutor seemed pleased with Wakil's capture. He also wanted the officers to search the residence. Excellent, now we've got cops working with prosecutors, with the ultimate objective of putting bad guys in jail. It was true counter-insurgency. And believe it or not, no one had to die from a missile strike by a Predator, in order to accomplish the task. No new enemies of America were developed in Dagu Village that day because of collateral damage.

● ● ●

I suggested we get Wakil off the scene for everyone's benefit. If some of his cronies came around, Wakil's presence would only inflame the situation. It was time for us to roll him back to the Counter-Terrorism Division for interrogation.

Agent M drove, while Sammy and I sat in the back with Wakil. I had to make sure Wakil was clear on a few points while I had his attention. I asked Sammy to convert to exact translation mode.

"Wakil, do you know why you're in handcuffs?"

"No, I'm wondering what the problem is."

"You know what the problem is. To save time, I'm going to get straight down to business. You've been building bombs. Those bombs have killed and wounded many people. You're going to jail for that, and we're going to make sure you don't get out this time. What I want to ensure is that the bombs stop. I know you have several people in your crew, and that other family members are involved. You're lucky because we are the police and not the military. But if one more bomb goes off on this route, I'm going to give the grid coordinates for your house to the U.S. military. They're not into arresting people. They'll just have a Predator shoot a missile and kill your whole family. I suggest that you get word back to your friends to stop, or your family will pay the price. We're not into killing people, but if I give them the coordinates, it's out of my hands. All those young commanders are just itching to kill someone and get a medal. Remember that."

Wakil was listening. I think he took it to heart because it was believable. Everything I told him was the truth and not a threat. I would do exactly what I told him. Now, I couldn't be sure anything would actually be done, but Wakil didn't know that. We would obviously continue our case and hopefully make some more arrests. However, the main objective was to just stop the killing.

We pulled into police headquarters and parked inside the counter-terrorism compound. We were met by several counter-terrorism officers, who escorted Wakil into one of the front rooms. We had arrived just in time for breakfast. Warm bread was already spread out on a vinyl tablecloth on the floor. Hot tea was being poured. A small skillet of scrambled eggs with chilies was steaming hot and looked delicious.

Here is the difference between the way we do business in the States and how business is done in Afghanistan. While Wakil was responsible for killing people, he was still a person. The Code of Pashtunwali is firm. Wakil was a guest and would be treated as such. The handcuffs were removed and he was expected to sit down and have breakfast with us. I asked one of the officers to take a photograph of the gathering. It would make a good picture for the first sergeant, who was responsible for kicking off this investigation. The bewildered look on Wakil's face tells a tale. There he was, eating breakfast with an American *crusader*, *an infidel, a non-believer,* he perceived. The same type of person a day earlier, he had been trying to kill. Now, a *crusader* was passing him the eggs and filling up his teacup. *What the hell was going on?* I was supposed to be his enemy. Why was I treating him like a brother? Wakil discovered that it's hard to hate a man while breaking bread with him at the same table. Human nature overcame hate, religion, politics, and negative, ill-informed perceptions, as Wakil slowly passed me the skillet of eggs.

After breakfast and tea, Sammy and I sat in on the interview conducted by the counter-terrorism officers. Wakil wasn't going to admit to anything. The only thing he explained was that a couple of years earlier, he had been picked up and accused of building bombs. They took him to the local district center, where he stayed for less than two days. While he was there, some Americans took his picture and his fingerprints. He was subsequently released without any charges. That experience, I believe, made him feel even more untouchable and confident about his actions.

I tried to give Wakil some friendly advice. I felt sorry for him. Not for him being in our custody, but for the physical pain he was potentially about to endure.

"Wakil, you may want to talk to us. What you don't realize is that in seventy-two hours, NDS (National Directorate of Security) is going to come over here and take custody of you. They don't play by the same rules we do. Think about that for a minute."

Wakil wasn't admitting to anything. It didn't matter, because we had more than enough evidence against him, already. He was out of the game and on his way to prison.

True to my word, NDS did eventually come and take custody of Wakil. Reportedly, after a few minutes of *interrogation*, Wakil began to sing like a bird. Now, whether or not he was giving accurate information is debatable. Wakil admitted to building bombs, planting bombs, and detonating bombs. Some of the bombs, we hadn't even heard about. I believe that he admitted to a bunch of things he wasn't involved in as well. He might have even told them where Jimmy Hoffa was buried and who really shot Kennedy. It is amazing what a person will say when you hit them enough times in the face with a rubber hose and squeeze their nuts with a set of pliers—not to mention the creative efforts of a person's mind, after a good session of surfing on the old waterboard.

I told Sammy I needed to roll to the prison camp to get out the preliminary report. I had to beat the military HUMINT (human intelligence) folks on the FOB. Their thirty-dollar sources would be calling, saying that NDS or some other unit had captured a bomb maker. I had to make sure our unit received the proper credit and that accurate information was reported. Sammy would roll back out to the scene and give me a report on any evidence seized. I told him that any evidence needed to come straight to me at the Paladin compound.

I had called the first sergeant and the major on their cell phones, but cell service inside the Paladin building was poor. I couldn't get them on the line to tell them the good news. Once I got back, I was finally able to share the success with my fellow gringos. The first sergeant was fired up, and rightfully so. It was his research that started the case.

An hour later, Sammy called to tell me they had seized a *bunch of shit* from Wakil's house. I asked him to have our guys bring the stuff straight to the prison camp. It needed to be analyzed at the lab. However, the prosecutor didn't want to give up the evidence, and insisted that it go with him. If he had no evidence, how could he convince the court to keep Wakil in jail? I had to extend my personal guarantee to the prosecutor, that I would bring back the evidence after it was analyzed. Luckily, he accepted the agreement.

I knew the young soldiers working the gate wouldn't take it lightly if they elected to search our vehicle, and found themselves staring at bomb making materials. Since I didn't feel like looking down the barrel of an M-4, I asked one of the EOD techs to accompany me to the gate. It would be better if a uniformed soldier was there to receive the men and the evidence. A sergeant first class (SFC) volunteered for the job and agreed to help me. While I trusted my people and their judgment, it was obviously smart to have an American EOD tech inspect the cargo, prior to bringing it onto an American prison camp.

...I would like to make a side note about this particular soldier. If I had asked for his permission to do so, he would have told me no, because he doesn't like to be in the spotlight. But I want to fill you in, because it's motivating and I consider this gentleman to be a true hero. I had known him for several weeks before someone made the point that he had a prosthetic leg. I thought they were referring to someone else, because the guy never let on that he was any different in physical capability. Come to find out, he had been blown up in Iraq while on a mission. Although the blast had only affected his foot, he made the doctors take off his leg, up to the point to where he could wear a prosthetic limb and stay in the army. He was nowhere near ready to quit and wasn't going to let the injury slow him down. There he was, serving in Afghanistan like any other soldier. He was a true warrior. So, if you are currently deployed in some far away land while reading this, just remember that your situation can always be worse. Remember this gentleman who continues to serve, minus his leg, and without complaint.

We met the agents at the front gate and escorted them to the area where the clearing barrels were located. The gate guards, as suspected, got nervous once we opened the back hatch. As they approached, I stepped out of the way to let military talk to military.

"What's going on?" the young soldier asked.

"I'm a sergeant first class with EOD, and this is a Task Force Paladin issue. It's alright, I've got it."

"I need to know what's going on," he insisted, as his eyes focused on the rolls of wire, and not the rank on the SFC's uniform.

"No you don't. I already told you I've got this. Go back to your post," the SFC said calmly.

The SFC inspected the cargo and gave it the all clear. In the boxes, there were various types of remote control devices, tools, flashlights, and voltage meters. A plethora of wire was among the contents. It all amounted to a bomb maker's workshop. Not a bad haul for a day's work. It was a lot of good evidence the prosecutor could take to court. It would ensure that Wakil stayed in jail. (That's if he lived through the NDS interrogation.) As we attempted to make it through the final checkpoint, the gate was abruptly closed by the young soldiers. The SFC in charge of the entry control point came out and talked to the EOD SFC. Everything was eventually cleared up and we were allowed to pass. However, we knew we would be raked over the coals from then on, whenever we tried to enter the prison camp.

The items taken from Wakil's residence were eventually sent to Bagram for analysis. This particular case would make the rounds as a success story for how the system was supposed to work. It was a team effort, from how the initial evidence was collected by EOD personnel, to the lab personnel lifting the fingerprint, to the staff judge advocates generating the prosecutorial package, to us capturing Wakil, to the Afghan prosecutor putting Wakil in jail. The case involved many people and several programs. Because of cooperation and coordination, we were able to take a bomb building expert out of the equation.

*...**Thanks again to the EOD first sergeant** for connecting the dots and kicking off the case. Had it not been for him, Wakil and his crew would still be digging holes at night and killing people. His concern for the safety of the young soldiers set in motion a chain of events, which would ultimately save an untold number of lives.*

CHAPTER 8

The Threat is Out There

After an early morning case that resulted in us seizing a large cache full of ordnance, Sammy and I decided to head back to the prison camp and get something to eat. We had been up since 3:00 a.m., and had already done more than a day's work in the interest of justice. We dropped the vehicle off at the Paladin compound and headed toward the chow hall.

We were laughing, joking, and carrying on after a successful mission as we walked in. A young, fat, female soldier sat with her big ass on a stool at the entrance. As I started to scan my ID card, she spoke.

"He can't come in here," she said, while pointing at Sammy.

"What do you mean, he can't come in here? He's with me."

"He can't come in because he's wearing sandals. There's no open-toed shoes allowed in the chow hall."

The three of us looked down at his sandals at the same time. Sammy had forgotten to change into his tennis shoes that he usually wore while on the prison camp.

"Oh, I'm sorry about that. We just got in and he forgot to change. It won't happen again," I said, in an attempt to solve the problem diplomatically with the young vixen.

"He still can't come in here," she said nastily.

Are you *fucking* kidding me? In Afghanistan, the majority of men wear sandals. There we were, supposedly as guests in their country. Their country; where sandals are the norm. But this nasty little bitch was subtly belittling my friend who had been in harm's way all morning. I would try one more time to resolve this felonious infraction through diplomacy.

"Look, I understand that you're just doing your job, but he's with me and we need to eat. We've been up since three. Can you just make an exception this one time? I assure you it won't happen again."

As I was saying this, three young soldiers walked in wearing their physical fitness uniforms. They each had on a pair of shorts and a t-shirt. They appeared as if they had just left the gym after working out for three hours. They were saturated with sweat from head to toe. The scent of sweaty balls, ass, and putrid armpits wafted my nose as they brushed past and scanned their ID cards. It was quite acceptable for Americans to come in and eat while dripping sweat all over the salad bar, but it wasn't OK for my friend Sammy to wear open-toed shoes.

"I'll get my NCO so he can tell you the same thing," she said in disgust, while wrestling herself off the stool.

I guess sitting on your ass all day is a stressful job. I was pissed. She had embarrassed Sammy, who really didn't understand what he had done wrong. She had embarrassed me, because Americans are not supposed to act like that. I was embarrassed for her. She turned the corner, leaving her M-16 leaning up against the wall. It had a magazine in it. The rifle was left unsecured, in the company of a nasty old contractor, and a nasty old Afghan unworthy of being able to eat because of his sandals.

"Screw it, I'm going to take this bitch's rifle and teach her rude ass a lesson. Then we can debate whether or not security is more important than playing fashion police."

I reached to take her weapon but Sammy grabbed my hand and stopped me, thus saving me from my redneck ways.

"Sir, don't do that. It's not worth it."

"Sammy, you don't deserve to be treated like that."

"Sir, what do you always tell me? Stay under the radar. Don't bring attention to us or they'll shut us down. Let's go."

*...**Sammy was right**, and was the voice of reason. I was severely pissed off and would have stolen her rifle, had he not stopped me. (Of course, she could have retrieved it from the lost and found at the MP station.) That's how strongly I felt about the situation because it summed up why we, as Americans, had failed in Iraq and Afghanistan. It was because we had absolutely no respect for the locals. We referred to them as local nationals. People from other countries were called third country nationals, or TCNs. Both terms were utilized in a derogative sense. It amounted to segregating different classes of people. Don't think so? How about the fact that local nationals weren't supposed to use certain latrine facilities? How do you explain that? You might as well have put a sign next to the door that read, "Whites and Blacks Only." Whether we want to admit it or not, all of these small, minute issues, when combined, amount to segregation. When segregation takes hold, different*

classes of people emerge. Referring to people as "local nationals" de-humanizes them and suggests that they are inanimate objects. Objects aren't worthy of the same treatment or basic dignities as human beings. The culture fostered because of this, explains why it was acceptable for the young American girl to treat a "local national" with a total lack of respect. Had she perceived him to be Mr. Sammy, the person, the encounter would have ended differently. In her eyes, he didn't rate any respect and subsequently, neither did I.

I walked away from that chow hall disgusted with what had transpired. I decided to buy Sammy lunch out in town at our favorite restaurant. There was no stress there. No segregation. I was always treated as an equal by the employees, unlike the way Sammy had just been treated. They all knew I was an American. They knew I was not Muslim. But those were never factors in how I was treated by the Afghan people.

● ● ●

The colonel called and reminded me that we needed to refuel the vehicles. We told him to meet us at the prison camp in about thirty minutes. Back on FOB Fenty, we awaited their arrival so I could escort them to the fuel pumps. It was easier for me to meet them at the gate because the Americans always gave them a hard time. Today would be no different. I walked to the front gate as usual, but was told by the young soldiers that I needed to report to the control room. *Sure, no problem. What would be the issue this time?*

I walked back to the security building and found the control room. A small one-foot by one-foot wooden window was where you were instructed to knock for service. After knocking a few times, it was opened by a young soldier.

"Yes, they told me I needed to come back and talk to you gentlemen," I said, very diplomatically.

"Ah, yes sir. Why are you bringing in Afghans through the coalition lane?" was the question presented in an investigative manner.

"Because they're undercover police officers," I said.

"They still need to go through the local lane to be searched."

"No, actually they don't. They have security badges for the FOB. Their badges allow them to carry weapons on the FOB. They have two forms of police identification. They have police badges. They're enrolled in BATS. I've gone through months of paperwork to ensure that they can come and go from here, in accordance with policy. Please explain to me why they need to be searched. They're authorized to carry weapons, so what's the purpose?" I asked.

A young, slightly obese soldier was sitting at a desk against the wall. He turned toward me and got up from his desk. He had on a pair of glasses so thick, it made me dizzy when I tried to look him in the eye. It also made it difficult to realize he was actually as cross-eyed as the day is long. It was hard for me to believe he had passed either the physical fitness test or the vision test. Nevertheless, he was standing in front of me.

"Sir, don't you know?"

"Know what?" I asked in bewilderment.

"No one's told you? Don't you know?"

Was this a trick question? What the hell was he trying to say? Spit is out, my good man, because I'm confused.

"Sir, don't you know? The threat is out there!" he exclaimed, while pointing toward the front gate.

I initially thought he might be joking, but he had said it in such a serious manner. Now, he was just standing there staring at me through those thick glasses. His head was tilted way back for some reason, as if he was trying to get the right angle through the trifocals. If it wasn't for his neck, his head would have rolled off his shoulders.

I almost fell out. I just came from *out there* and didn't see the threat. Maybe he could be more specific or point it out. I must have missed it. I stood by for the rest of the story.

"Sir, the threat is out there. Don't you know those five FBI guys just got killed on that FBI base because they didn't search everyone? Why don't you want us to search the Afghans? They could be bringing in suicide vests and planning attacks. They could have an IED in their vehicle. The threat is out there!"

Wow. I was stunned. I really didn't know how to respond. First of all, this young soldier did not have his facts straight. What he was referring to was the seven CIA affiliated employees, who were killed on FOB Chapman months previously. The incident, in my opinion, was the result of not properly searching an informant.

It would have been useless to try and explain to this kid the difference between undercover police officers and informants. He had already been brainwashed by hours of intelligence briefs prior to his recent arrival in Afghanistan. Most of those briefs were probably presented by people who had never been to Afghanistan. He had been scared into believing that every Afghan was trying to kill him. Fear makes money, so the façade has to be maintained. I told myself to keep quiet, but I just couldn't help it. They were wasting my time, so I figured I'd waste theirs.

"Let me make one observation, my friend. If a green Ford Ranger pulls up to the gate with guys wearing ANP uniforms, do you search them?" I asked.

"Well, no. They're Afghan police," he conceded.

"Do you ask for ID?"

"No, they're in uniform," he answered.

"OK, my guys are undercover. However, they have every ID card required. So, why do you have to give them problems every time?"

It was no use. They could not get over the possibility that guys wearing man-jams could be trustworthy. They were all *Hajjis* (term used improperly by American personnel in a derogative manner, to refer to Afghans) and therefore could not be trusted. I walked away from the window twice as pissed as I was earlier at the chow hall. Imagine how the Afghans must feel when dealing with these attitudes. However, they have no recourse and just have to endure whatever the gringos dictate.

I still had to refuel five vehicles and get them ready to roll for the next operation. So here's what I was forced to do: I had to walk out through the gate, cross the street, and get a vehicle from my agents. I had to drive back in, go through the backscatter (x-ray) machine, drive over to the pumps, and refuel the vehicle. Then, I had to return it outside the gate. It took me four hours of shuttling vehicles and waiting in line at the pumps to get us back into a mission ready status. It should have taken twenty minutes. Not to mention the fact that I got my nuts fried by the backscatter van, after making a total of seven trips onto the prison camp that day. I would always yell at the kid in the backscatter van to turn it off, but I know he never did. I wonder about the health ramifications I'm going to eventually suffer due to all that extra, unnecessary radiation I was enduring.

● ● ●

Let me elaborate a bit further on what the young cross-eyed kid had brought up. This will not sit right with many, but it needs to be said. The seven CIA folks who died at FOB Chapman did not have to die. If you are a family member of those who died, my heart goes out to you. If I were you, I would be mad as hell. I would not settle for the excuse that Director Panetta put out as to why they died. They did not die because they screwed up personally. They died because of internal politics, bureaucracy, and the lack of commonsense culture which is the norm within our intelligence community.

They died as heroes doing our nation's dirty work, but this particular incident was simply avoidable. While I was not there, and do not believe in quarterbacking others who make decisions in the heat of battle, this was a different situation. It was not a decision made in the heat of battle. This was a planned meeting. There were no exigent circumstances. It was planned. That's my first point.

Even a rookie cop is taught to search suspects. A rookie narcotics agent is taught to search informants. Why? It is because neither can be trusted one hundred percent of the time. Informants often have ulterior motives for cooperating with government agents. This is police work 101. If you don't properly search a guy, he will stash a gun or dope in the back seat of your police car. Ask any cop and they will confirm this for you. It has probably happened to most of them, and that's where they really learned the lesson—through real life experience. You're taught this in a classroom but a real life experience makes you a believer.

So, how did a guy blow himself up and kill seven people? It would be very easy for me to believe the guy killed one person. I would chalk it up as a risk of being in this line of work, but deductive reasoning says that seven dead agents are indicative commonsense was ignored. But why? I'll share with you my theory.

First of all, this particular informant, a Mr. Humam Khalil Abu-Mulal al-Balawi, claimed to have information on Al-Qaeda's number two man, Ayman al-Zawahiri. The mere possibility that it was true would have put a lot of focus on this particular informant. I would suspect that hundreds of people from ten different agencies bickered, argued, and jockeyed to see who would get to talk to the guy and control him. During the turf fight, there were numerous PowerPoint presentations and operational plans drawn up as to how this gentleman would be handled. By the time he made it to FOB Chapman, agents on the ground probably had strict instructions as to what to do, what to say, how to search the guy, what *approach* to use, etc. They probably had a hundred people sitting in cubicles in D.C. pulling their strings, and telling them exactly how things had to be done. It got them killed because the ability to use commonsense had been taken away from them.

It is possible that the agents weren't allowed to leave FOB Chapman because after all, *the threat is out there*. OK, no problem. They should have sent a couple of trusted Afghans to pick this guy up—Afghans who knew what the hell they were doing. I guarantee you, if I had sent any one of my crew to pick the guy up, this would not have happened. Once the guy was in our vehicle, he would have already been clear. The worst that could have happened was that he took out one or two of my people, during the initial meet and greet. That's it. That's the secret of how this incident could have been avoided. Search your informants. If you don't, you might die. If a rookie narcotics agent knows this and abides by it, how could seven agents of the U.S. Central Intelligence Agency not know?

Director Panetta authored a three-page letter in reference to this incident. It offers no explanation, but it details steps he has taken since.

> Message from the Director: Lessons from Khowst
>
> Last December, our Agency family lost seven courageous and talented colleagues in a terrorist attack at Forward Operating Base Chapman in Khowst, Afghanistan. These dedicated men and women were assigned to CIA's top priority—disrupting and dismantling al-Qa'ida and its militant allies. That work carries, by its very nature, significant risk. CIA is conducting the most aggressive counterterrorism operations in our history, a mission we are pursuing with a level of determination worthy of our fallen heroes. We will sustain that momentum and, whenever possible, intensify our pursuit. We will continue to fight for a safer America.
>
> Earlier this year, I directed that a task force of seasoned Agency professionals conduct a review of the Khowst attack. The purpose was to examine what happened, what lessons were learned, and what steps should be taken to prevent such incidents in the future. In addition, I asked Ambassador Thomas Pickering and Charlie Allen, a highly accomplished former Agency officer, to conduct an independent study of the Khowst attack and to review the work of the task force. They concurred with its findings. One of CIA's greatest strengths is our ability to learn from experience, refine our methods, and adapt to the shifting tactics of America's enemies.
>
> The review is now complete, and I would like to thank those who participated. They did our Agency a great service. It was, to be sure, a difficult task—especially since key insights perished with those we lost. Perfect visibility into all that contributed to the attack is therefore impossible. But based on an exhaustive

examination of the available information, we have a firm understanding of what our Agency could have done better. In keeping with past practice, we will provide the Khowst report to the Office of Inspector General.

In highly sensitive, complex counterterrorism operations, our officers must often deal with dangerous people in situations involving a high degree of ambiguity and risk. The task force noted that the Khowst assailant fit the description of someone who could offer us access to some of our most vicious enemies. He had already provided information that was independently verified. The decision to meet him at the Khowst base—with the objective of gaining additional intelligence on high priority terrorist targets—was the product of consultations between Headquarters and the field. He had confirmed access within extremist circles, making a covert relationship with him—if he was acting in good faith—potentially very productive. But he had not rejected his terrorist roots. He was, in fact, a brutal murderer.

Mitigating the risk inherent in intelligence operations, especially the most sensitive ones, is essential to success. In this case, the task force determined that the Khowst assailant was not fully vetted and that sufficient security precautions were not taken. These missteps occurred because of shortcomings across several Agency components in areas including communications, documentation, and management oversight. Coupled with a powerful drive to disrupt al-Qa'ida, these factors contributed to the tragedy at Khowst. Each played an important role; none was more important than the others. Based on the findings of the task force and the independent review, responsibility cannot be assigned to any particular individual or group. Rather, it was the intense determination to accomplish the mission that influenced the judgments that were made.

There are no guarantees in the dangerous work of counterterrorism, but the task force identified six key areas that deserve greater focus as we carry out that vital mission. We will:

- Enforce greater discipline in communications, ensuring that key guidance, operational facts, and judgments are conveyed and clearly flagged in formal channels.
- Strengthen our attention to counterintelligence concerns while maintaining a wartime footing.

- Apply the skills and experience of senior officers more effectively in sensitive cases.
- Require greater standardization of security procedures.
- More carefully manage information sharing with other intelligence services.
- Maintain our high operational tempo against terrorist targets, even as we make adjustments to how we conduct our essential mission.

I have approved 23 specific actions recommended by the task force, some of which I ordered implemented months ago. They provide for organizational and resource changes, communications improvements, tightened security procedures, more focused training, and reinforced counterintelligence practices. These include:

- Establishing a War Zone Board made up of senior officers from several components and chaired by the Director of the National Clandestine Service. It will conduct a baseline review of our staffing, training, security, and resources in the most dangerous areas where we operate.
- Assembling a select surge cadre of veteran officers who will lend their expertise to our most critical counterterrorism operations.
- Creating an NCS Deputy within the Counterterrorism Center, who will report to the Director of the Counterterrorism Center and ensure a more integrated effort across Agency offices.
- Conducting a thorough review of our security measures and applying even more rigorous standards at all our facilities.
- Expanding our training effort for both managers and officers on hostile environments and counterintelligence challenges.
- Creating an integrated counterintelligence vetting cell within our Counterterrorism Center that focuses on high-risk/high-gain assets, evaluates potential threats, assesses "lessons learned," and applies the latest technology and best practices to counterterrorism operations.
- Designating a senior officer to ensure that all the recommendations are indeed implemented.

We've now taken a hard look at what happened and what needed to be done after the tragedy at Khowst. While we cannot eliminate all of the risks involved in fighting a war, we can and will do a better job of protecting our officers. Drawing on the work of the task force and its insights, it's time to move forward. Nothing in the report can relieve the pain of losing our seven fallen colleagues. By putting their lives on the line to pursue our nation's terrorist enemies, they taught us what bravery is all about. It is that legacy that we will always remember in our hearts.

Leon E. Panetta

The highlights are that the director appointed a task force to complete a review of the incident. The task force identified six key areas that deserve greater focus. While one of them is to *require greater standardization of security procedures*, there are none that simply say, *search your informants*.

To continue with the dog and pony show, the director indicates that he has approved twenty-three specific actions recommended by the task force. Some include creating a War Zone Board, creating new positions, and a new cell (unit). It equates to spending millions, if not billions of U.S. taxpayer dollars, which doesn't even address the issue at hand. Smoke and mirrors are used to cover up reality.

...*Search your informants. I give the CIA that for free.* *I'm not a scholar, they will say, so how can an uneducated man be taken seriously? Well, I went an entire year meeting informants from within the Taliban, with a team of Afghans. Not on the prison camps, but out in the streets and hills of Afghanistan. Most of the time, the informants knew they were meeting with an American. I was constantly a target. But not one of my agents got hurt and I'm still alive to generate this manuscript. That trumps all the knowledge and education possessed by the decision makers within the CIA. It speaks for itself—res ipsa.*

A secondary suggestion is to cut out the hundreds of middlemen, analysts, scholars, and politicians, who insist on weighing in on a situation, and let the guys on the ground handle business by applying commonsense. Policy and bureaucracy should never trump commonsense.

To the families of the lost, I apologize if this line of thinking is brash. However, no one in the federal government listens until someone dies. That's the way it is. Some good people died and I'm still reading nothing but nonsense. If I were you, I would file lawsuits on behalf of every member of your family affected by the loss. Do not settle out of court. Take this all the way to a jury trial. If you do not, others will die because of bureaucracy and incompetence within the CIA. I will voluntarily testify as a subject matter expert on your behalf. But you don't need me—just subpoena any cop in America. They will all testify the same on the subject of searching people. Do not settle for the answers given to you by the U.S. government. Government officials do, and will lie, to save face. The way they get away with it is hiding behind the excuse that everything is classified, or secret. It's only secret because people want to keep their high-paying jobs. They don't want to be exposed to the truth that they may be incompetent at what they're supposed to be doing. And just maybe, the taxpayers are paying them too much.

All this letter tells me is that Director Panetta is playing politics and hoping that memories fade as time passes. *The threat is out there,* only if you allow it to be.

CHAPTER 9

Permission to Engage

In March, Jim called and broke the news that I was going to be getting a partner, at least temporarily. I was to take the guy under my wing and get him ready to take over the team in Laghman Province. This worried me. In the contracting business, you never know what type of guy is going to come down the pipe next. He could be a hero, a coward, a worker, lazy, or anywhere in between.

My first thoughts were this person could pose a threat to the way I did business. If he turned out to be some retired FBI agent who had spent thirty years in a cubicle abiding by the rules, he could be trouble. I wasn't exactly abiding by any policies other than our own, so a spy in the mix could potentially shut down my operation. I would have to put on a show for the guy while he was in Jalalabad, and as soon as possible, I would ship him out. I decided to voice my opposition to Jim.

"Ah, yeah. About this new guy coming down here. If you have somewhere else for him to go, feel free to send him there. I'm just fine on my own and really don't want a partner. To be completely honest with you, I don't need a rat or a coward screwing up our operations."

"We know, and we've got two guys to choose from. We're going to talk to them and see who would be the good fit. I know why you're worried. Just relax, and I'll get back with you."

"OK, but warn whomever you're planning on sending that if he comes here, he had better be ready to work. If he's not, don't send him."

Sammy and I discussed the situation and agreed we would not trust the new guy with what we had going on. We would just show him around the prison camp and maybe bring a couple of our agents to meet him. We wouldn't take him outside the camp. A couple of days later, Jim called back with an update on the situation.

"Man, have we got the perfect guy for you," said Jim.

"Why do you say that?" I asked.

"Just trust me. You're going to like this guy. That's all I'm going to say. His name's Paul Cuellar, a gang investigator from Phoenix. You're going to love him. He's coming with his interpreter as well."

"Alright, I'll give him a chance. But he'd better be ready to work."

At least the guy wasn't a retired Fed. Aside from DEA and Border Patrol agents, they're about as useful as the hair on my tits when it comes to catching real bad guys.

● ● ●

Paul Cuellar and Nazar got off the helicopter in a swirl of dust. Sammy and I were standing by to greet them. Instead of 5.11 wear, Paul was sporting a blue pair of Dickies pants, typical of gang attire from back on the block. He was only missing a red bandana. Paul was Mexican-American, which was a good thing. His darker skin and features would allow him to blend into the Afghan population with ease. Within minutes, it was blatantly apparent Paul and I were going to get along just fine. I can't quite describe it, but we were both on the same sheet of music from the beginning. He was one of the good guys who could be trusted.

After introducing ourselves, we drove back to the Paladin compound and helped Paul and Nazar settle into their rooms.

Over chow, we got to know one another. Nazar was originally from Mazar-i-Sharif and was Hazara. His features, typical of the Hazara people, made him appear Asian. Nazar had spent many years working as a local interpreter for some Special Ops guys, since the early years of the conflict. He had seen plenty of combat and knew how to survive. He had recently received his paperwork to go to the U.S. and had moved his family to North Carolina. After a couple of months slinging vegetable crates for minimum wage, he decided it was in his best interest to go back to Afghanistan and make the big money. His Pashtu and Dari were both excellent. I liked him from the start.

We had to come up with a nickname for Nazar and I was itching to throw in my vote.

"My friend, I think we'll call you Jackie Chan!"

Nazar laughed, as did everyone else at the table.

"OK, OK. Jackie Chan," Nazar said, while grinning from ear to ear.

The name stuck. It made sense, because he looked more Asian than he did Hazara. Looking back, I'm not real sure Nazar actually knew who

Jackie Chan was at the time. He was just laughing because everyone else was laughing.

Paul was from Arizona and had been working street gangs near Phoenix. He was on a leave of absence from the police department and was also a captain in the army reserves. While he looked like a typical thug in the hood, I was surprised to learn Paul had a master's degree. He was obviously a very intelligent guy. He also had a laid back personality.

Paul and I decided to break free from Sammy and Nazar so we could talk, cop to cop. My initial plan of keeping the new guy in the dark was immediately scrapped. The sooner I could get Paul spun up, the sooner we could get back to work. Over the course of the afternoon, I explained to him exactly how I was operating and the reasons for keeping a low profile.

"The first thing we've got to do is get you and Nazar some man-jams. We've got a barbecue scheduled for tomorrow. It will be a perfect time for you two to meet everyone."

With that, we rummaged through boxes in the shipping container and found some man-jams that would work, at least temporarily. Paul had a muscular build, so all of the clothes were a bit too small. A top was all he really needed to get out to the safe house, anyway. He and Nazar could get some man-jams tailor-made on the prison camp, as soon as we had free time.

● ● ●

The next morning, Paul, Jackie Chan, Mr. G, and I loaded up some goodies for the barbecue and filled the hatch with bottled water. I had also invited a guy named Carlos to come with us. Carlos tried to decline by saying he was too busy, but I insisted he come. Carlos worked in the regular law enforcement program and had been instrumental in the initial formation of my team. He was a good guy, but he had spent his entire career at the FBI. This career path made him unsuitable for undercover work or running operations in Afghanistan. He was just too regimented. He needed permission to take a piss. I would have thought of him as a total stiff, but he had recently been caught doing something that totally redeemed his status with me. Old Carlos got stone cold busted, tagging one of the female docs over at the hospital. *Hell yeah! My man!* At least someone was getting laid. That made him OK in my book. Even though he was outwardly the most policy-driven individual, subliminally he was saying *screw you* to General Order Number 1, as well. It was probably the only thing he had ever done wrong in his life, but you just can't screw with nature. Unfortunately, due to his lack of discretion and secrecy, the whole affair had gone public. He was ratted out to country management by a lazy bible-thumper who did nothing his entire tour but complain, sleep, and get paid. Carlos didn't want to leave Afghanistan, but after three years of service he was on his way out the door. I felt it an honor to have him come to the barbecue prior to his departure. I would not have taken no for an answer.

When we arrived at the safe house, I introduced Paul and Nazar to the crew. Since Nazar was a native Afghan, he obviously fit right in. All of the team members welcomed Paul and Nazar with open arms. It was as if they had been there for months.

"Paul, we work hard, but we party even harder."

With that, Sammy came rolling in with about five cases of Heineken in the trunk of his Corolla. The cases were immediately raided. Paul just started laughing.

All he could say was, "My man…"

Imagine this scene: Muslims, Baptists, and Catholics, hammering down warm beer like it was cool, in a Muslim country. After about thirty minutes, I made the command decision to deploy Sammy back to the bazaar for a re-supply. Sammy was too drunk to drive and Agent S volunteered to make the run. When it comes to drinking beer, you can't be messing around. Any one of us could catch a bullet or get blown up on the next mission. Therefore, I could not allow the beer to run dry during a get-together. That was a serious issue with me.

Lieutenant Colonel N came rolling in with a couple of sheep in the back of a Forerunner. Paul laughed at the sheep because they looked as if they were just riding shotgun. Nazar laughed at Paul, for laughing at the sheep. Lieutenant Colonel N had also recruited our favorite butcher, so that everyone could concentrate on partying. Ever tried to kill and dress a sheep when you're drunk? I don't recommend it. The hatch of the Forerunner was opened, and the sheep were allowed to run loose in the compound while the butcher set up the barbecue pit.

As the beer was consumed, the language and cultural barriers broke down even further. Drunks understand other drunks, no matter what language is being spoken. It's a scientific fact. I should know, because I've spent twenty years of my life directly involved in the research.

The butcher eventually announced he was ready to kill the sheep and we all gathered around to enjoy the show. Lieutenant Colonel N held the sheep down on its side, while the butcher sliced halfway through its throat with a razor-sharp knife. The blood spewed about two feet in the air, as if he had cut into a water hose under pressure. I drunkenly complained that the butcher almost stained my two-year-old shower shoes I was wearing. As the sheep kicked its hind legs in a futile attempt to run out of the situation, the blood continued to leak from the open gash in its neck. The leak eventually turned into a slow ooze and the sheep stopped kicking. We took this opportunity to take a prized photograph of the situation.

The butcher cut the head completely off and we strung the sheep's body from a doorway. He immediately began to break down the sheep with efficiency. While this was a common event for us, it would be a shocking process to the majority of Americans.

While we were standing around watching the butcher work and getting drunker by the minute, I broke out a small boom box I kept in the Forerunner. Kid Rock's music soon filled the air inside the compound. I was the only one singing along, but heads were definitely bobbing to the beat. Except for the small fact there were no women present at the party, we were rocking it out, Afghanistan style. I wasn't paying attention, but the local mullah had commenced the call to prayer. No less than eight beers were immediately abandoned on the hood of the nearby Forerunner, in consecutive *thunks*. Agent S grabbed the beer from my hand and set it down on the hood, while waving his hand back in forth. He was signaling me to cease the partying for a minute. The butcher was pushing buttons on my boom box and finally got the music stopped. There was a pause in the party while the mullah completed his chant.

Allllahhhhhhhhhhhh Akbar! Allllahhhhhhhhhhhh Akbar!

The second the mullah finally stopped, the party was back on. Nine beers were grabbed off the hood and the drinking continued. The butcher hit the play button on the boom box.

This made me laugh. We were already violating every principle of Islam and breaking several of the laws of Afghanistan. That was acceptable. However, it was NOT OK to party during the call to prayer. That's taboo and is going just a wee bit too far. It was a point duly noted and would always be adhered to during future get-togethers.

I decided it was time to break out the red plastic cups and a recently acquired bottle of Jim Beam. A buddy in Kabul had bought the bottle for me at the U.S. embassy. My old lady had sent some more bottles of lime and lemon juice in the last package, so we mixed up Jim Beam and

Coke over ice. Once everyone had their drink mixed to their liking, I drunkenly proposed a toast.

"I'd like to propose a toast to our new friends, Paul and Nazar. May Allah keep them safe during their time with us."

With that, it was bottoms up. The Jim Beam went down smooth. All of my Afghan brothers loved American whiskey. The only hard liquor they could typically get was from Pakistan. It was cheap home brew from who knows where. I drank it on many occasions even though it tasted like rubbing alcohol, and probably was. On this occasion, everyone was drinking the good stuff.

I realized I had left out someone very important in the toast.

"I'd like to propose another toast. I want to drink to that no good, low down, dirty *motherfucker* back in Washington who came up with General Order Number 1. Screw that son of a bitch!"

Me and Paul laughed. Everyone else just drank. Sammy was getting too drunk to translate, anyway. I continued my passionate rant about General Order Number 1, until I fell off the porch and spilled most of my drink on my man-jams.

The butcher was finally ready with the barbecue and the Kalahi, which is a dish made with lamb and some kind of red sauce. I have no idea what all is actually in it. I just know that it's good eating. A spread that would rival any Thanksgiving dinner was laid out on the table. We all ate until we could eat no more. At the conclusion of the meal, the Colonel and I took turns saying a few words. I made sure to recognize Carlos for his efforts and contribution in forming the team, which had become the number one producing police unit in the entire country. Cuban cigars were broken out and served as the perfect conclusion to the evening's festivities.

● ● ●

On the ride back to the prison camp, I let Paul drive because I wasn't capable. I was tore up and Mr. G was out of control. He would not shut up about something Sammy had said. I couldn't even tell you what the issue was. We strongly counseled him during the brief trip, to shut the hell up when we rolled past the gate guards. He finally complied and we made it through with no problems. Then, he fired back up to the point that it wasn't wise to roll straight to the compound.

We had to drive laps around the airfield, stop and get several Red Bull energy drinks from a local shop, and wait until he was through running his mouth. We laughed so hard at his silliness that it was impossible to try and talk sense into him. No reason to get frustrated with our brother—just drive another lap. This type of craziness is to be expected after a night of partying. It's the fiber that builds camaraderie. If no drama occurred at one of my parties, then it wasn't a real party.

● ● ●

The next afternoon after I recovered from a hangover, Paul and I discussed the best plan of action to get the Laghman team up and running. There was a prison camp in Mehtar Lam, which was the main city in Laghman Province. There were a couple of law enforcement advisors there, who were embedded with the military unit. They were supposed to be keeping in contact with the Laghman team until Paul arrived, but hadn't accomplished anything.

We were concerned that if Paul moved to Mehtar Lam, the battlespace renter would lock him down on the prison camp and not allow him to operate. That would kill the team. A way around it was for his team to operate out of Jalalabad. We had enough room to run two teams out of my safe house, and Mehtar Lam wasn't that far away. We discussed the pros and cons of the concept. Over the next few days as we debated what we were going to do, word got around to the battlespace renter that we were considering moving *their* team to Jalalabad. They had apparently been using what members the team had, as an intelligence collection agency. But, they weren't doing any proactive police work, which was their purpose and mandate.

A nasty e-mail was generated by the staff judge advocate who happened to be a state prosecutor back in the civilian world. The e-mail began with his introduction of something about being in charge of a safe neighborhood task force in Connecticut, and so on, and so on. We both laughed until we hurt, at that claim to fame. When you're addressing two veteran undercover guys who have done battle in the streets against real crooks and drug traffickers, don't try to posture. Connecticut? Are you kidding me? Has there ever been a crime committed up there? I don't even know where Connecticut is on a map. Is there really a need for a safe neighborhood task force in Connecticut?

The e-mail basically opposed us moving the team. What the guy didn't realize, was that we didn't work for him. Our teams didn't work for him, nor did they report their actions to him. For gringos like him, that's a hard pill to swallow but it is reality. Now we had petty politics involved in our planning. All it would accomplish was to slow down Paul's team from going operational.

"How do you want to handle this?" I asked Paul.

"Well, for politics' sake, it looks like we've got to go up there and have a meeting with these cats."

"Up to you, my friend."

"Yeah, he said to come on up whenever I could get a Molson flight. What do you think?" Paul asked.

"Here's what I think. You're right, we've got to go up there and address these guys, but we're not flying. That sets the wrong precedent. We'll just show up at the front gate."

"Sounds good."

With that, we arranged a meeting with the SJA, the sergeant major, the S-2 (intelligence officer), and the law enforcement advisors. We would hash things out in person.

When the day rolled around, Paul, Nazar, Sammy, and I headed toward Mehtar Lam. It was only about a forty-five-minute ride from Jalalabad. As we pulled up to the gate, Paul did the talking. The young soldiers were caught off guard when Paul began speaking perfect English. Since everyone in the vehicle appeared to be Afghan, they had us stage in the parking area until they could get someone to come and vouch for us.

"You think we should change into gringo clothes for the meeting?"

"Nah, screw it. I'm tired of changing clothes to appease the gringos. If they don't like my man-jams, I'll sit in the car," was my response.

After a few minutes, we were escorted onto the prison camp. Sammy, being a local national, was *persona non grata* on this trip. He had to stay with our vehicle out in the parking area. I apologized to Sammy for the inconvenience, but reminded him he really didn't want to be in the meeting, anyway. Sammy retreated to the Forerunner to take a nap.

We walked to the conference room where we were introduced to the sergeant major. He was a gruff old guy, but I liked him. I just got the feeling he was a no-bullshit type of person who got things done. He could be a friend.

We were then introduced to Major N, the rule of law coordinator. My initial impression was that he was an Ivy Leaguer who felt above the crew sitting in front of him. It could be the mere fact he's an attorney and is just programmed to act that way. A soldier from the battalion intelligence shop (S-2) was present, along with the two Company X law enforcement advisors.

I had planned on keeping my mouth shut and letting Paul do all the negotiating. As the discussions ensued, it was apparent Major N thought he was in some way in charge of the Mehtar Lam ISU team. I had to interject and explain to him, that a proactive police unit in a sovereign country does not answer to the U.S. military. He couldn't seem to comprehend the concept, in spite of the fact he's an attorney. I will spare all of the details of the meeting and get to the pertinent points.

"You mean you roll out with no one but Afghans?" was the question presented to me.

"That's correct," I responded.

"Without any coalition forces present?"

"If there were any coalition forces present, we wouldn't get anything accomplished," I said, matter-of-factly.

"Well, I think you trust these people too much," Major N insisted.

"I don't think you trust them enough," I shot back.

What I really wanted to say, though at the time I could not, was that I wasn't naïve for trusting *these people* and that *these people* were my friends. I trusted them with my life on a daily basis. At any time, they could have knocked me out and sold me to the highest bidder in Miram Shah, but they wouldn't. What I wanted to explain was that on many occasions, *these people* had held my head up, while I puked all over a Turkish shitter from drinking too much Pakistani home brew. Or, that I had taken care of them in the same manner. I knew their families. I had been a guest in their homes. We had done battle together.

I would rather be around *these people,* instead of sitting in this meeting, listening to what amounted to bureaucracy from an attorney. It was bureaucracy that served no purpose other than to impede our operations, which were actually producing tangible results and saving lives. I had to remain silent and let him think he was important.

"If you're operating without coalition forces, can you guarantee me that there won't be a friendly fire incident?"

Now, what kind of question is that? I can't predict nor guarantee what the American military will do. They kill innocent civilians all the time. If Paul and I were on the scene, we could guarantee that we would call and tell their operations center where we were and what we were doing. That was for our safety. I pondered why he was wasting our time over something that didn't concern him. He was a rule of law advisor, not the battle captain in the tactical operations center (TOC). He was basically a paper-pusher who attended meetings.

"I can guarantee you if we're there, your operations center will know about it, basically for our safety. Our biggest fear is the Americans, not the Taliban. The main thing we're afraid of is U.S. aircraft. Therefore, we'll make sure your operations center knows where we are."

This wasn't good enough for him and he continued to talk about operational issues he really knew nothing of. He was talking to the two of us like we were a couple of boy scouts. Having an attorney from Connecticut lecture you on how to run operations was comical, yet very painful. Trust me. We both wanted to laugh, but couldn't. The saving grace was I knew the *neighborhood warrior* couldn't talk forever, because either the chow hall or the gym would be about to close and he would eventually have to go.

The meeting finally came to an end and we began to depart. As we walked out, we were introduced to the battalion commander. I felt better about the situation as I got a read on the gentleman. He was a good guy and would be an asset. He didn't try to posture in front of us, like Major N had been doing. He was a real soldier. He seemed as if he was there to accomplish a mission and make a difference.

"How do you think it went?" I asked Paul.

"I think it went pretty good, but it looks like I'm going to have to move up here to appease these guys."

"Yeah, I agree. That elitist Major N thinks he knows it all, but he's out of his league. The comical thing is that he'll be back in suburbia in a few months, wearing his coat and tie and telling war-stories, and we'll still be conducting business."

"My man," was Paul's reply.

● ● ●

A few days later, Paul got a call from Major A, the commander of the Mehtar Lam team. A snitch had called about some guys planning a rocket attack against FOB Mehtar Lam. They had reportedly staged a couple of 107mm rockets in the hills outside of town, in anticipation of an attack that night. We needed to roll immediately and link up with the team.

"Hey man, I'm going to crank off a quick e-mail to all the battle captains at brigade, and let them know we're rolling out that way," I told Paul, hurriedly.

"Good idea. I'll call the sergeant major up there and let him know what's going down," Paul replied.

With that exchange, I fired off an e-mail to all of the battle captains at brigade. I cc'd a few more people, who may have had a need to know. I cc'd even more people, who may have wanted to know. I added a few more, who I thought would potentially cry, if they didn't know.

> Gentlemen,
>
> ISU Laghman is conducting an investigation of a possible cache located in Alingar District. Just wanted to give you a heads up that an operation will possibly be conducted this afternoon to seize the cache.
>
> SGM W***** with TF Iron Gray was notified of the preliminary information. We will update SGM W***** direct of any developments.
>
> We will provide an exact grid of the cache once it becomes available.
>
> If you have any questions, please contact myself at 079*******, or Paul Cuellar at 079*******.
>
> Respectfully Submitted,
>
> Mark Blackard
> ISU-Team Jalalabad

Paul got in touch with the sergeant major via cell phone and personally briefed him of the operation. The sergeant major told Paul he was on his way to the TOC and would brief everyone there. Paul would keep in close contact with the sergeant major as the op progressed and provide him with updates. We covered all of our bases on this particular case, so no one would be complaining once we got back.

"That should cover our six and appease Major N."

"Yeah, I think we're good."

"Let's roll."

With that, Paul, Nazar, Sammy, and I hauled ass toward Mehtar Lam and linked up with the team. Major A had already rounded up some uniformed officers to provide backup. Once a briefing was conducted, we rolled toward the target area. The ride took us up through several small villages. Farmers worked in their fields while children tended sheep.

Women washed clothes in the stream. We had ringside seats to a beautiful panoramic view of the countryside.

We made quite a spectacle due to the number of vehicles transiting through, and curious villagers rushed out to see what the commotion was all about. I felt no threat whatsoever from the people in the area. They were folks who were merely trying to make a living and live their lives.

As this was a good photo opportunity, I asked Sammy to get in the sunroof and take some video instead of functioning as the usual turret gunner. As he was filming, the children transformed into instant Americans. They were mesmerized by the camera and all wanted a piece of the action. Various antics were displayed in an attempt to briefly star in the footage.

As we continued to travel on the winding dirt road, which ascended the small mountain on its right side, I saw in the distance what appeared to be an antenna from an MRAP. Paul saw it as well.

"Is that an MRAP?"

"Looks like it."

"Yep, that's a convoy of gringos in front of us."

We had rolled up on a lumbering convoy of MRAPs having difficulty transiting the narrow roadway. We realized that one of the MRAPs belonged to the EOD guys. The thought crossed both of our minds, that the soldiers had somehow received the same information about the rockets and were en route to the same location. It would make for an awkward situation to be responding to the same location the Americans were. We needed to rule that out.

"Hey brother, call the sergeant major and see where these cats are off to," I suggested to Paul.

"Yeah, I agree my man."

Paul spoke to the sergeant major and learned the convoy was en route to a different location. Over the ridgeline from where we were going, a patrol had been hit with a roadside bomb. Five American soldiers were injured and an MRAP was disabled. Casualty evacuation (CASEVAC) operations were underway. The convoy in front of us was responding to the incident with a wrecker and an EOD crew to work the scene. It had nothing to do with our case. We all had a brief moment of silence, because we knew there were five young men across the mountain who were hurting. Hopefully, they would be alright. I thought about their families back in the States, going about their day, unaware of the tragedy which had just occurred.

We eventually came to a Y in the road. The MRAPs continued straight, to get to the far side of the mountain. We turned right, to get to the near side. We were soon back to making good time after being slowed to a snail's pace by the MRAPs. As we topped a small knoll, we emerged out onto the valley floor. To our left was the target area which was at the foot of the mountain in question. We all looked up to see a pair of F-15 fighter jets cutting racetracks. While they were there providing close air support to the patrol hit by the roadside bomb, it appeared as if they were there to provide air cover for us, on the near side of the mountain. Any potential bad guy would think they were our assets and would be stupid to try and hit us on the way out. We discussed the benefits of having the free, impromptu air cover. They would be our guardian angels and not even know it.

To our right was an open valley which was flanked by hills. We parked at the foot of the mountain and dismounted. Paul immediately made a call to the sergeant major in the tactical operations center and gave him our location. The sergeant major advised he had us on the board and would remain in the TOC to serve as our liaison during the mission. As the team began to hump up the mountain in an attempt to locate the 107mm rockets, my sixth sense kicked in for some reason.

"Hey Paul, I'll be up there in a minute. I'm going to call Task Force Paladin and give them our location as well."

I called and spoke to the duty officer. Since there were aircraft already on station, I requested that he forward our position report to the brigade. By doing this, we had double tapped the Americans that friendly troops were on the ground. Both the local battlespace renter and his boss (the brigade) had been advised of our grid. But just in case, I instructed an agent to deploy aircraft marker panels on our vehicle. The panels are basically huge pieces of reflective material, orange in color, much like a reflective vest someone on a road crew would wear. The agent deployed the panels on the hood of the Forerunner as directed.

I asked the ANP officers to crank their trucks and activate the strobe lights. Aside from popping smoke, there was nothing else that needed to be done. I returned to the crew as they hunted through the rocks for the cache. After several minutes of searching, one of the agents located our quarry. It turned out to be a field expedient (homemade) rocket launch kit, complete with two, 107mm Chinese rockets. There were fuses and timers, along with a spool of wire. It was everything the bad guys needed to launch those puppies at the Mehtar Lam base, later that night.

...The bad guys typically just lean the rockets up against a pile of rocks. They eyeball the direction, estimate the distance, and adjust the elevation of the rockets by hand. Most of the time, they can secure a hit somewhere on the FOB. They are proficient, because they had a decade of practice at the Russians' expense. Over time, they identified the best launch sites and avenues of escape. Experience pays. What's even more on their side is our stupidity. Why? Because all we did was move into the same areas the Russians used.

Since the local EOD unit was obviously tied up with the incident across the mountain, we called Ralph and his team.

"No problem," was the response.

Ralph and his crew would be there to assist us in about an hour. While we waited for Ralph, we took a few photos and chatted to kill time. We had good security set and we held the high ground in the rocks. Anyone would be committing suicide if they rolled up on us. With the F-15s overhead, we weren't worried about much.

About twenty minutes later, the F-15s disappeared. A single B-1 bomber with its black silhouette arrived to take their place. Everyone found the B-1 fascinating. I personally had never been out when a B-1 was on station. It began to cut racetracks in what seemed like slow motion when compared to the F-15s. It was entertaining and somewhat peaceful to watch—almost hypnotizing. We felt a continued sense of security due to the free air cover. The young ANP officers thought Paul and I had called in the air support. We saw no reason to tell them the truth.

Ralph's convoy arrived in a cloud of dust and quickly went to work. They made sure the rockets weren't booby-trapped and gave the all clear. Photos were taken for the report and then Ralph's people collected the rockets for disposal. After chatting for a while, we all departed in one huge convoy as the sun started to set.

● ● ●

Back on the prison camp, I congratulated Paul on a job well done. It was their first case in the midst of building the logistical framework of the Mehtar Lam Team. After grabbing some free food at the chow hall, we sat down at our computers to knock out what should have been a quick report.

As I logged in and stared at my in-box, the short-lived celebration of the day's events came to an immediate halt. An e-mail was waiting from our favorite major. I didn't know whether to get pissed off, or bust out laughing at the stupidity it contained. Either way, its existence meant that we had problems to address in the political realm. *What an asshole*, I thought to myself. Did the guy really not have anything better to do? Didn't he have some type of work to do as the rule of law coordinator? Oh well, let the games begin.

> Gentlemen,
>
> 1. Our TOC did not know anything about this operation. The TOC SGM and Battle Captain are not pleased. This is exactly what I was talking about during our meeting.
> 2. Our scout platoon just hit an IED in the area where you want to operate in. Grid XD********. They have five injuries and conducting MEDEVAC operations.
> LTC E*** will no doubt be discussing the above with MW6.
>
> Major ***** * N****
> Rule of Law Coordinator

As I read the e-mail, the first thing I wondered was why Major N didn't pick up his cell phone and call us, if he had an issue. Why would he send an e-mail over SIPR to people who were already out in the field? My Forerunner isn't equipped with a SIPR drop. The part about *LTC E*** will no doubt be discussing this with MW6 (brigade commander)*, made me laugh. What was he going to discuss? The fact that we interdicted a rocket attack on the camp he lived on? That we possibly saved the life or limb of a young kid serving his country? They were all heinous crimes, no doubt.

Damn attorneys. They screw up everything in America. Now we had one trying to screw up things halfway around the world as well.

"Paul, come read this crap. This guy is an idiot. Did he not read the part that says we had notified the sergeant major? He's an attorney for Christ's sake."

As Paul read the e-mail, "That's ridiculous. Let me call the sergeant major and have him handle this."

Paul made a quick call and advised everything would be taken care of. It still didn't set my mind at ease.

"I told the sergeant major and he was pissed. The sergeant major was in the operations center the entire time. He said not to worry or respond and he would handle it."

"That's good news. But I told you, every time we do an op, someone is going to cry about it. Every time."

True to his word, the sergeant major squared things away with the major and the battalion commander. But, not before another round of critical e-mails came full circle. Again, no thanks or praise for preventing a couple of 107mm rockets from slamming into the Mehtar Lam gym or chow hall.

The next e-mail pinged back began as follows:

MAJ N****,
I appreciate your righteous indignation...

I will spare you the rest, but I found that part comical. Our operation apparently disrespected Major N and his crew sitting safely on the FOB, and caused him *righteous indignation*. I don't even know what the hell that means. I think it means we pissed him off. I may have mentioned, I'm not a scholar.

• • •

A few days later, Paul decided to roll to Mehtar Lam with Nazar. He wanted to snoop around and see what kind of accommodations they could come up with, before they permanently moved. We met up later that night, and he filled me in on a conversation he had while in the Mehtar Lam operations center.

"My man, you ain't going to believe this shit," chuckled Paul. He then began to laugh uncontrollably.

"What happened, dog?"

It took a few minutes for him to get calmed down enough to tell the story. Apparently, he had rolled into the operations center to inquire if the sergeant major was around. The young soldier manning the radio made a call to another office, during which, the phone was on speaker. Paul could hear as the young soldier tried to explain to the person on the other end about his (Paul's) affiliation.

"Oh, you mean those crazy civilians who roll around undercover in soft-skinned vics?" asked the voice on the other end.

"Yeah, those guys," was the soldier's response.

Paul continued on with his story, "But that's not the funny part. After he got off the phone he said, 'Man, you guys don't know how close you came to dying the other day.' I said, what do you mean? He said, 'The F-15s had to leave because they had to refuel, so the B-1 came on station. About five minutes later, the B-1 pilot calls over the radio and says, *Be advised, I've got multiple civilians on the ground, heavily armed... Permission to engage!* We checked the board and realized it was you guys. We told him, *Negative. Negative. Negative.* He comes back on again and says, *Are you sure? I've got multiple civilians, heavily armed, moving toward friendly positions. Request permission to engage!* We told him again that it was a negative and eventually got him called off.' My man, I couldn't believe what the kid was telling me."

"Are you kidding me?" I asked in disbelief.

"I'm not kidding you. It's a good thing we had comms with the sergeant major. We almost got smoked."

Ten seconds of serious silence between Paul and I were followed by a round of laughter that brought tears to our eyes. I was in the floor trying to catch my breath without pissing my pants. It had to be one of the funniest things I'd ever heard in my entire life. It was especially funny, since I was still living to enjoy the tale. After we regained our composure, we pondered how the B-1 crew members could see *multiple civilians, heavily armed,* but could not see aircraft marker panels and blue lights flashing. I wondered if the excitement of getting to kill some people turned to tunnel vision. How else could you see people on the ground but not the police vehicles nearby?

I reflected on the moments I was staring up at the B-1 and thinking how peaceful and graceful it looked—not knowing that simultaneously, the graceful bird was plotting to rain hell down on us. Some may find the incident disturbing, but to Paul and I it was just another adventure. Cheers to the pilots and the weapon systems officers (WSOs), wherever they are right now. I love those guys. There have been many beers consumed in their honor for attempting to send us to paradise. That's an adventure even Richard Branson can't buy.

CHAPTER 10

Hitchhikers and Cellos

Mr. G began to lay the groundwork that one of his family members was getting married in Kabul. There would be a huge party the night before, and the wedding would be quite a fiesta in itself. According to Mr. G, his family was more liberal than most Afghan families in Kabul. The women would be in the same room as the men during the wedding. There would be dancing, pretty girls, booze, etc. Now, I really didn't believe any of this because I had already stayed at his family's house for a couple of nights. I would agree they were more liberal than most families, but they would not outwardly display that modernism in a public place. I was certain there would be many pretty girls there but no mixed dancing. The booze would be underground, as usual.

Mr. G wanted to go in a bad sort of way. It would be in essence, a huge family reunion for him. He was sugar-coating things to try and get me interested. He said everyone was invited.

"Mr. G, don't make any promises to your family right now, but I'll try to get you there, at least for one night. I can't guarantee you'll make both events."

I told Paul about the invitation and suggested we go. While it wouldn't be a blowout, it would be something to do. Besides, a nice weekend in Kabul was in order. We both needed to get away from the gringos and their silliness for a while. Therefore, we solidified the trip. I made the decision to not take Sammy. While Sammy was my right hand man, he was a bit too conservative to roll to a wedding that was potentially mixed. I would have to ease him into that aspect of western culture, later on.

As the weekend approached, we gave word to Mr. G and Nazar that it was a go. We would roll to Kabul on Friday afternoon. The big party was Saturday night and the wedding was on Sunday night. We would play it by ear as to when we needed to come back. On Friday, we loaded up into a Toyota Hilux pickup and headed west. We linked up with Major A and another agent from the Mehtar Lam team, just outside of Jalalabad. Major A was rolling in his Toyota Corolla. We decided to take both vehicles, as there was no need to be cramped. The ride was beautiful as usual, and uneventful. A few hours later, we were pulling up to Mr. G's aunt's house in Kabul. We enjoyed a good meal and were treated like kings. Soon, Paul, Nazar, and I were trying to get some sleep. The guest room was nice. It had a television with satellite, and the typical Afghan mats lined the walls. The only thing that wasn't nice was the shitter. It was as if a modern home had been constructed around a one hundred year-old outhouse. It made no sense.

Since there were several women living in the home, Paul, Nazar, Major A, and I had to be escorted to the shitter by one of the males of the house. It ensured the females would be out of sight and sound when we needed to roll out and do our business. Even though we were houseguests and good friends, we were not family. Only family can be around the females as a general rule. It wasn't a huge deal, but I'm getting old. It sucks to have to wake someone up in the middle of the night to go take a piss, especially on multiple occasions. It's also an Afghan custom for men to hold hands with other men, whom they consider to be good friends. So, I reluctantly had to engage in some moonlit walks to the outhouse, hand in hand with my good buddy Dawood. That's one cultural aspect I could never get accustomed to and frequently voiced my opinion about. I'm OK with hugging other men. It's no different than hugging your brother or your father, but holding hands with other men is going too far. Again, I have nothing against my homosexual friends out there. You're all welcome to come over for dinner, anytime. However, I'm not going to hold hands with you.

It just gave me a queasy feeling every time I found myself in that situation. To be honest, I really don't like holding hands with a woman, either. It's just too romantic and feminine. I would typically act as if my cell phone was ringing and use it as an excuse to break contact with whoever had grabbed my hand. In the middle of the night, during the short walk to the pisser, that story wouldn't fly.

The next morning, I was exhausted because of Paul and Nazar and their sawing of logs during the night. As we ate breakfast, Paul and I discussed our plan of action. Did we really want to go to the wedding?

It meant we would have to endure a lot of formalities over a period of several hours. While there should be booze, I wasn't looking forward to dancing with a bunch of men—especially since I knew there would be a plethora of women on the other side that I couldn't even look at. What else could we get into?

●●●

After we finished eating breakfast, Paul, Nazar, Major A, and I fired up the Corolla and headed out. Our first stop was a small convenience store located on Jalalabad Road. Major A knew the owner. We found ourselves sitting on milk crates in the back storage room, drinking semi-cool Heineken in cans. We laughed at our current situation. No women and no whiskey. At least we had beer. But, we seriously needed to find some *ho's*.

The conversation turned to telling stories about our past adventures in chasing women. A discussion ensued as to the observation that Afghanistan's night life was like one big gay bar. I've been to several gay bars during my travels—it was merely an objective observation, without any prejudice or bias involved. The problem was, the four of us were heterosexual by nature and loved the ladies. Several beers later, Nazar began making calls on his cell phone.

"Hey guys, I'm talking to my good friend in Mazar, and he says he knows some girls up there that will come over."

"What the hell are you talking about?"

"He knows some bitches up there that will party. Plus, he's got a hookup on beer and stiff drinks. He wants us to come up there and hang out because we haven't seen each other in years."

"How well do you know this cat?"

"He's one of my best friends. I've known him all my life."

"No way he's got some ho's up there."

"If he says he does, then it's one hundred percent."

"Tell him you'll call him back in a few minutes."

That was too good to be true. You mean that Nazar's buddy could have a ready-made party waiting for us? All we had to do was roll? This had to be evaluated because Mazar-i-Sharif wasn't exactly a couple of miles down the road. It's located in the northern part of the country, just south of Uzbekistan.

"Bro, how far is Maz from Kabul?" I asked.

"I think it's about ten hours," replied Nazar.

"What's the route like, going up there?"

"I don't know. It's been a long time since I traveled that road," replied Nazar.

Paul and I discussed the pros and cons of rolling to Mazar-i-Sharif. If Nazar's friend was telling the truth, then it would obviously be the better time. But, it was a haul to get up there. The discussion ping-ponged back and forth as to whether or not to go. A couple of more Heinekens were sent down the hatch by each of us.

We discussed the fact we would have to roll back to Mr. G's crib and pick up the Hilux. It was four-wheel drive with a diesel engine. More importantly, it had all of our gear in it. It was packed down with bottled water, food, our change of clothes, GPS units, my good camera, AK-47s, and everything else we typically traveled with. I don't carry a gun when I plan to drink heavily (for good reason), so the only weapons we had with us at the time were Paul's 9mm and Major A's Makarov. It would take us an additional hour to pick up the Hilux because we were already close to the road that cut north.

I had on my person, my military common access card and $300 U.S. dollars. That was it. Paul had some cash and so did Nazar. As far as funds went, we were more than OK. The half-drunken debate raged on for a good twenty minutes.

"Here's my position on this. I ain't down with another night of holding hands with Dawood every time I've got to piss. I love that young brother, but that shit ain't happening tonight," I offered.

"Yeah, and I ain't down with dancing with a bunch of men at the wedding," Paul responded.

Nazar sat anxiously and waited for Paul and me to make a decision because his buddy was standing by for a call back. I caught myself staring down into my beer can in deep thought. Something came over me and I realized we were bickering over the obvious. It was obvious what the responsible and sensible answer was. We were crazy to even contemplate driving three hundred miles through Taliban held areas, without our equipment and weapons, just to party with some chicks. If something broke bad, we would definitely be in hot water with our company and the gringos. Or worse, we could get our asses handed to us because we weren't properly armed. The responsible decision had to be made. There was just no other choice. We shouldn't go.

But, I wasn't sensible, responsible, nor the type of person to ever turn down an adventure.

"Fuck it, let's go," I said, with a serious tone. I then chugged the last half of that warm Heineken.

The trip was on! Laughter broke and the other three followed my lead. As we got up off those dusty milk crates, I knew we were in for a good time. We loaded up a trash bag full of Heinekens for the road. As I fired up the Corolla, I realized that beer and cash was all we had packed for a cross-country journey through no-man's land. Talk about a wild feeling of freedom.

As I turned left, off of Jalalabad Road, an older Afghan police officer was standing on the corner as if he were waiting on a ride.

"Hey Nazar, ask the comandante if he needs a ride!" I said, while pulling over in front of the gentleman.

The old gentleman was pleasantly surprised at the offer and was soon sitting in the back seat between Nazar and Major A. As everyone in the car was three-quarters of the way drunk with beers in hand, he wasn't going to be left out.

...During the last get-together *at the safe house, Nazar had offered Major A a taste of his small bottle of Jack Daniel's, as a gesture of good will. To Nazar's surprise and dismay, Major A turned the bottle upside down and killed it (drank all of it) in one swoop. When Major A handed him back the empty bottle (with a smile on his face), Nazar could do nothing but look at it with sadness in his heart. "But, we were supposed to share," was the thought going through Nazar's mind at the time.*

Major A pulled out a small metal flask from the inside of his vest. He had filled it with Captain Morgan's Rum, which he had obtained by hoarding several of the miniature bottles my wife had sent. As a goodwill gesture, Major A handed it to the comandante so he could have a sip of Americana. The comandante took the flask, held it up as if he was saying *cheers*, and said a quick thanks to Major A. The comandante then turned the flask upside down and killed it in one swoop. As he finished off the last drop, he let out an *ahhhh*. Major A's eyes were as big as saucers, as the old comandante handed him back the flask. Major A was in a state of temporary shock. The rest of us, including the comandante, just laughed. Nazar explained to Major A that what goes around, comes around.

The comandante cracked open a Heineken, chugged it down like it was water, and then went for a second. With the windows down and the radio thumping to Indian music, the five of us rolled north like kings. As we drove, the comandante explained he was in charge of the road we were on, from Kabul to an area about an hour north. If we had any problems whatsoever, we should call his cell phone and he would take care of us. No one would be permitted to give us any problems. We had his personal guarantee on that. We thanked him for the gesture.

After about thirty minutes of traveling, we came upon a checkpoint. For some odd reason, a heavy-set female ANP officer was part of the crew manning the post. She was the first one to approach the car. Her eyes lit up when she saw the big boss riding bitch (seated in the center) in the back seat, with three hillbilly Pashtuns and one Hazara in the car. After the comandante had a brief conversation with her, the Dari speakers in the car all laughed. Apparently, she was startled as we drove up, because it looked like we had kidnapped the guy. Why else would he be rolling with these hillbillies? It was a good laugh, but the comandante decided he should probably get out and get back to work. His wisdom told him if he rolled with us, his wife would kill him by the time he made it home. We shook hands and hugged the old gentleman. He would be a friend from then on. A couple of beers, jokes, and hoopin' and hollerin' about women built a strong bond during a simple thirty-minute ride.

● ● ●

We continued north until the turnoff road for Bagram Air Base. We were rolling at light speed, when Major A brought up the fact we might not be on the best road to where we needed to go. I stopped the car at a roadside stand, and we queried an old gentleman and his grandson. According to them, we could take the road we were on, but it turned to crap in about an hour. It would be rough going. But, if we backtracked a bit, there was a new road we could take. It was much better. We thanked him and decided to take his advice. Soon, we were on the new road trudging north.

We left Kabul Province and rolled into Parwan Province. A brief passing through Kapisa Province and Panjshir Province, brought us back into Parwan. As we rolled through the Salang Pass, the mountains became increasingly beautiful. We stopped to take pictures with the mountains in the background. Several hundred feet below us, we could see a small dump truck with several men walking around it. We realized they were shoveling snow into the back of it to take down the mountain and sell.

As we departed our photo site, Nazar and Major A began to discuss the Salang Tunnel. I had never heard of it before. They told us the tunnel had been built by the Russians, and that many people had been killed inside of it over the years. That didn't sound good. We would soon find out how scary it would be.

The entrance looked like a half circle. The type to where, if big trucks didn't drive in the middle of the road, they would scrape the ceiling. As we entered the tunnel, the road immediately turned to mush. It was dark, and water dripped from various crevices in the ceiling. As we passed a truck coming toward us, it kicked mud up on the windshield. I turned on the windshield wipers to clear the grit and only succeeded in making matters worse. The road had decomposed completely through years of neglect. It was part asphalt, but mostly just mud. The chunks of asphalt jutted up like small icebergs amid the mud and the gravel. Mounted on the ceiling were a few turbine-looking fans that were obviously designed to be part of the original ventilation system. It appeared as if they hadn't worked in decades. Supposedly, during the fighting between the Taliban and the Northern Alliance, the ventilation system was damaged and the entrances were bombed. Either way, there was no ventilation currently inside the tunnel. We all began to hold our breath, due to the amount of carbon monoxide fumes that permeated the air. I looked over at Paul, who had his t-shirt pulled over his nose and mouth. This was definitely the tunnel of death.

"Gents, if the car breaks down, get out and start running. If we stay in here much longer we're going to die."

I wasn't joking. If there were a crash inside the tunnel, everyone would suffocate. We continued to trudge ahead. I had to utilize all of the wit I had left in me, to negotiate the jagged mounds of asphalt. After twenty minutes in the gas chamber, we still had no idea when the ride would end.

"My man, I'm getting dizzy," Paul complained.

Everyone was getting light-headed but we finally reached freedom. We rolled the windows down on the Corolla and let the cold, crisp air flow in. While it was a relief, it wasn't enough. I had to pull over and get out for a while. I think we all came close to actually passing out.

● ● ●

After a short break, I had my wits back and was ready to continue the journey. We rolled into Baghlan Province and were making good time. At a roadside stand, we picked up some ice-cold Red Bull energy drinks that had been chilled with mountain water. The road conditions went from excellent to terrible, and back again to excellent. There were no herds of sheep to get in our way, like in other areas. As we rounded a corner on the far side of a small town, we ran straight into an Afghan police checkpoint. The young Afghan stopped us with a level of alertness as he gripped his AK-47. I rolled the window down.

"As-Salamu Alaykum, andewal," I said.

My friendly greeting in Pashtu was met by a flurry of Dari I didn't understand. The officer opened the door to the Corolla and motioned for me to get out. I stood face to face with him as he began to question me. I understood that he was asking something, but couldn't quite figure out what he wanted to know.

I introduced myself and told him my Afghan name by saying, "Zemanu Shahuddin de Tajuddin Zoyem." Then I said, "Ze Mazar-i-Sharif tazem," which informed him I was on my way to Mazar-i-Sharif.

The officer responded in Dari by raising his voice and asking the same questions. It was obvious he was getting pissed. I quickly exhausted the rest of my Pashtu and found myself out of words. I began to think something was wrong, because I had transited a hundred different checkpoints before, with no problems. Still about one-quarter drunk, I decided to give up the ghost and show him my identification card.

"American, my friend. Sorry, but I don't speak any Dari. Nazar, get your ass out and talk to this gentleman."

The three amigos inside the Corolla were having a good laugh at my expense, listening to me struggle through the encounter. Looking back, I wonder why Nazar or Major A didn't get out and intervene. They must have been enjoying the entertainment, and were probably taking bets to see if this guy was going to jack me up. They finally got out and began explaining to the officer I was indeed a gringo, and that I didn't speak Dari. He didn't believe them and called out the commander. The commander didn't believe them either, and closely inspected my ID card. I said a few words in English to demonstrate my heritage, and Nazar told him we were with OGA (other government agency), which all the Afghans interpreted to be the CIA. It was kind of funny that we always posed as CIA agents, to cover the fact we were undercover police officers.

"Why didn't you guys stop at the last checkpoint?" he asked.

"We did stop at the last checkpoint," replied Nazar.

"They called and said a Corolla blew through the checkpoint and refused to stop."

"Comandante, we stopped. It wasn't us. If we didn't stop at the last checkpoint, then why would be stop at this one?"

Nazar had made a good point. The comandante had no choice but to believe him out of logic, and because ninety percent of the cars in Afghanistan are Toyota Corollas. That was a vague description at best. The old comandante still had a puzzled look on his face.

"Where are you all going? Are you lost?" he asked.

"We're going to Mazar-i-Sharif," Nazar told him, while we all pointed in the direction we were heading.

"But…you can't go that way," he said.

"Why not, is there a bridge out?"

"No, there's Taliban that way. They're all along that route. You'd be crazy to go that way at this time of night. There have been several attacks recently. We don't even go through there after dark. It's too dangerous."

Nazar translated what he had said. Paul and I started laughing.

"You mean there's Taliban that way?" Paul asked jokingly, while pointing down the darkened road.

"Hell, we've been looking for them *motherfuckers* all day. That's exactly where we're going! You sure that's the right way? We don't want to miss them," I added.

More laughter erupted from our crew. It's hard to keep a straight face when you get stopped by the police at a checkpoint, while you're less than sober.

"I strongly advise you guys to not go that way. You should spend the night here, and continue your journey in the morning when the sun comes up. If something happens, we're not coming up there until day break," the comandante sternly warned.

"Thanks for the advice sir, but we've got somewhere we've got to be," replied Nazar.

The old man shook his head from side to side, half in amazement and half in disbelief. I think he really had a hard time with the fact there were two Americans in the mix of this unusual set of circumstances. He turned and walked back into the building, while still shaking his head. He had done all he could do and was washing his hands of the situation. We were free to go.

● ● ●

Back on the road, it was time to make haste. As we got closer to Mazar-i-Sharif, the road turned pristine, as if it had just been paved. I stepped on the gas and we were cruising at around ninety miles an hour. The roads in Afghanistan are strange. It's as if after they are constructed, there is absolutely no maintenance. Due to this, road conditions are subject to change drastically, without warning. As I was commenting to Paul about how nice the road was, I saw a darkened bar across the pavement coming up in front of us. I jammed on the breaks and got us down to about sixty before the impact.

BA-BAMMMM!!!! The little Corolla launched and bucked. It sounded as if the front end had been torn out from under us. We had hit a speed bump at about sixty miles per hour, while hard braking. It had to have been eight inches high. The car swerved back and forth as I regained control. Nazar and Major A woke up in the back seat, scared to death.

"Holy shit! Who in the hell would put a speed bump out in the middle of an interstate?" I yelled.

Sure enough, there had been an asphalt speed bump erected in the middle of a long straightaway. It served entirely no purpose. Occasionally, shop owners will erect speed bumps to slow the traffic in front of their roadside stands. While inconvenient, I understand the concept because it serves a purpose. There was no one within miles of this straightaway through the valley floor. I didn't slow down, because the logical explanation was that it had been put there by the bad guys. The Corolla continued on as if nothing had happened. After this trip was over, I would conclude that Corollas were the toughest vehicles on the face of the earth, with the Forerunner coming in a close second. Nazar and Major A eventually calmed down and went back to sleep, while Paul and I chatted.

"Hey man, I meant to tell you some funny shit that happened the other day," Paul began. "Me, Nazar, and old Major A were on the way up to Mehtar Lam when we ran into a checkpoint. This young cat didn't know us and starts poking his nose in the ride, looking around. Major A pulls back his fishing vest, where the cat can see his pistol and magazines. He even had a couple of frags hanging off his shoulder rig. The young ANP kid's eyes got big as hell, and he just froze. Major A says, *Dog, I don't know what you're looking for...but we got it all!* We laughed, and the kid just waved us through."

Paul kept telling funny stories to keep me awake and alert for any more monster speed bumps. Our laughter woke up Nazar, who joined us in the storytelling. He chimed in from the back seat with old Dari proverbs that somehow applied to our current and past adventures. The only one I remember is, *If you see a donkey walking down the street and it's not yours, don't stop it.* That basically means to mind your own business, but in a more entertaining way. The rest escape me but are definitely words to live by.

We eventually passed through a police checkpoint on the outskirts of Mazar-i-Sharif and were back into friendly territory. We also passed by the airfield that was being run by the Germans.

The first thing I noticed which made Mazar much different from Jalalabad, was the fact everyone in the city had electricity. When we got close to Mazar, it seemed like every little mud hut had a porch light. The electric grid came down from Tajikistan, therefore everyone in the area had good, reliable power.

Nazar called his buddy to get directions to his crib, and soon we were in a neighborhood with dirt streets. We finally arrived after a grueling twelve-hour road trip packed with danger and good times.

● ● ●

Nazar and his friend embraced, as it had been years since they last met. I attempted to drive the Corolla into the qalat, but the dirt ramp was too steep and it bottomed out. It took the group to push it off the dirt and back out into the street. In the process, it broke the muffler a bit. No problem. Nazar's friend ordered his little brother to spend the night in the car to ensure it was safe. I felt bad for the kid, but that's the way things are done there. If an elder tells you to do something, you do it. There's no negotiating. Just get the job done. He climbed into the Corolla, while the rest of us went inside.

The house itself was inside a typical qalat and was built out of brick. There was sufficient lighting and it was very modern for Afghanistan. When I asked where the bathroom was located, it was the same situation. There was an outhouse built into the front wall—no light and no running water. Just go out there in the darkness and shit down a hole. I still don't understand it because the house had a nice, modern kitchen with running water. Again, as I've said before, it's as if they build a house, and when it's finished, they realize they forgot the bathroom. *Ah, just put an outhouse over there. That'll do.* At least I didn't have to hold hands with Dawood when I wanted to piss. Nazar's friend had dispatched all the women (who actually lived with him) to a relative's house, in anticipation of our visit. Therefore, we didn't need a personal escort to the shitter.

As our trip had taken longer than we expected, the girls who had come over to party had already gone home. Nazar's buddy would call them back the next day. Tonight, we would just get some sleep and relax.

The guest room was similar to the one at Mr. G's aunt's crib. I staked out a mat and was soon fast asleep.

The next morning, we awoke to a feast being served by Nazar's friends. As guests, we weren't allowed to raise a finger. We were treated like royalty. After the meal, it was time to take a trip into town to wash up at a public bathhouse. Reportedly, they had hot water. We could pick up a couple of cheap towels and a bar of soap in the bazaar.

The place turned out to be a large facility. Steam filled the air as men filed in and out. Nazar gave the keeper the money, which amounted to about fifty cents each. Inside, there were individual stalls. I closed the door behind me and began to strip down. There was a bucket placed beneath a faucet that was dribbling steaming hot water. At the rate it was flowing, I quickly deduced that I only had the one bucket of water to work with. There was a small bowl floating in the bucket of water. You would use the bowl to splash water on yourself. I lathered up and used every last bit of that hot water to rinse off with. Feeling refreshed, I donned my man-jams and abruptly realized they were emitting the strong odor of sweat, fermented ass, and two-day-old beer.

We arrived back at the house to find that a beer run had taken place. A couple of bottles of bootleg liquor had also been acquired. The glue on the label was still wet and *Jack Daniel's* was misspelled—no telling what was really in those bottles. The festivities began with Nazar's friends breaking out some musical instruments. A cello of sorts had been constructed out of a piece of wood and a burned-out gas can. It only had a few strings, but actually sounded pretty good. There was also a homemade two-string guitar. We were soon rocking out to this two-man band, waiting on the girls to show up. About twenty minutes later, Nazar's friend received a call. One of our guests had arrived. We looked out the front window and watched as a blue *burqa* entered the qalat.

"Damn that *bitch* is fine!" I said, and Paul agreed.

Of course, we had no idea what she looked like, but the *burqa* indicated she was a woman. That was a one hundred percent improvement from what we had a few moments earlier. She walked in, and took off her *burqa* in the extra room. Shyly, she walked into the guest room and quietly sat down.

She was a beautiful girl, about thirty years old, with long dark hair. Underneath the *burqa,* she was wearing a tight pair of jeans and a chic blouse. If you didn't know any better, she could have been any chick, pulled from any bar in America. After getting a couple of drinks in the girl, she was part of the crew. Now we had the beginning of a real party. More people would show up as the time passed. The afternoon turned to evening, and evening into night. I'll spare the complete details, but it was a great time and created some unforgettable memories.

The next afternoon after we recovered from hangovers, it was time to roll. A debate ensued as to whether or not we should wait until the next morning to go. Paul brought up the fact that he didn't want to roll back through the tunnel of death during rush hour. It was suicide. We'd already taken a year off our lives on the trip up. I agreed with Paul. We should leave to where we hit the tunnel of death at about two in the morning. There would be little to no traffic at that time of night. We thanked Nazar's buddies for a great time and headed south.

We got to the edge of the first mountain and began our ascent. The fog thickened as we climbed, to the point you couldn't see five feet in front of the car. Nazar was driving and wanted to pull over. Major A agreed. They argued it was just too much of a risk to try and navigate in those conditions.

"Good idea, Nazar," I said, from the back seat with Paul. "Pull over."

Once Nazar pulled over, Paul and I got out. I opened the driver's door and Paul opened the front passenger door.

"Now get out and get in the back seat. We'll take it from here."

Not exactly what Nazar had in mind when he pulled over, but we had been screwing off long enough and needed to get back to Kabul. With the windows down, Paul hung out the passenger's side while I hung out the driver's side. Paul would tell me when I was getting too close to the edge of death, and I would turn back left. Through teamwork, we trudged forward at about five miles per hour. We knew once we cleared the mountain, it would be smooth sailing. Nazar and Major A continued to express their disapproval. They weren't ready to go to heaven, just yet.

We approached the Salang Tunnel at around 2:00 a.m., as planned. There were only a few vehicles on the road. I stopped briefly, so that everyone could hyperventilate and get some oxygen in their lungs. Paul pulled his t-shirt back over his head, and off we went. I drove as fast as I could and we made good time. It wasn't near as bad as the previous transit, mainly because we weren't having to fight traffic.

● ● ●

As the sun came up over Kabul, we were rolling in with the morning rush. We picked up Mr. G and headed east, toward Nangarhar. Mr. G admitted the wedding sucked. Hundreds of women, but you couldn't talk to any of them. There were men dancing with men. It had been a real hoot-nanny. It affirmed our decision to go north.

While writing the words for this adventure, I had to look at a map on the Internet to help recall our exact route. I discovered that reportedly, in February of 2010, about two months prior to our trip, there had been a series of major avalanches at the Salang Tunnel. Around one hundred fifty people died from either asphyxiation or hypothermia. There were hundreds more injured. Further history: In 1982, a fire broke out after a fuel tanker crashed inside the tunnel. The fire reportedly killed sixty-four Soviet troops and over one hundred Afghans. Avalanches supposedly killed people in 2002 and 2009 as well.

Who knew? As Mullah Paul would put it later on, *Crazy stuff my man. Always one step ahead of death is how cats like us live. Cheating death at every corner, or in this case, at every tunnel.*

CHAPTER 11

Operation Night Train

On April 14th, at around 7:00 p.m., one of our agents got a call from the same snitch who had given us the 3,395 pounds of hash back in December. It was the snitch I always referred to as *The Professor*. The Professor informed us that a mule train loaded down with hash had left Hesarak District in Nangarhar Province. It was to travel through Deh Bala District and onto Kot District. The ultimate destination of the hash was obviously Pakistan, around the Peshawar area.

According to The Professor, he knew the mule train had to pass by a T intersection, on a particular trail in Kot District, sometime during the night. After the mule train passed that particular intersection, he would be unsure as to which way they would go. Either route would take them into Pakistan. If we wanted to hit it, we had to get to the T in the trail as soon as possible or the opportunity would pass.

The Professor also advised the mule train was being escorted by armed men who would travel well ahead of it. If any contact was made by the armed guards out front, the men driving the train would turn the mules and flee in the direction from which they came. The Professor suspected the men actually driving the mules would be armed as well.

Sammy relayed the information to me and told me to get to the safe house. As I was loading up the Forerunner, Paul and Nazar came in from a meeting in Mehtar Lam. They were wearing their new, pristine, white man-jams. They looked like Afghan bankers.

"We've got to roll! There may be a gig going down tonight so we've got to haul ass," I yelled, as they got out of their car in front of the Task Force Paladin compound.

They jumped in and I apologized for putting them on the spot on such short notice, but I definitely needed them on this operation. I briefed them on the hash deal back in December and the information The Professor had provided. Basically, we were going to set up an ambush a few miles from the Pakistan border, and rip another load of hash from Taliban drug traffickers.

At the safe house, an operational briefing was conducted with the team. The basics of the deal were laid out. We were to link up with two informants on the road to Torkham Gate. They would guide us to a vehicle staging area out in the middle of nowhere, and then take us to the T intersection in the trail, also out in the middle of nowhere. There, we would ambush the mule train and seize the dope. A quick getaway was in order, to avoid any conflict. It was a simple operation, as long as we could quietly get in place before they passed by the choke point. If they passed by before we got there, we would we lying in ambush waiting on ghosts. Time was against us.

"OK, let's go get with the general and hopefully get a raid team rounded up in an hour," I said, and headed toward our vehicle. The colonel stopped Sammy and asked him a question in Pashtu.

"Sir, the colonel wants to know what we are going to do with the mules," Sammy inquired.

"What do you mean?" I asked, as I pondered how it had anything to do with the gig.

"Well, according to Afghan law, we have to seize the mules. If we don't seize the mules, the prosecutor will want to know why. It's an issue."

"It's not an issue. Besides, how in the hell are we going to get those mules out of there? After we rip that dope we may stir up a hornet's nest. I can't shit five horse trailers, Sammy."

"Sir, I know, but everyone is worried about it. The courts will want to know what happened to the mules."

"Tell him we'll cut the mules loose. They'll find their way home. They know their way better than we do."

With that, several other agents got involved in the conversation over what we were going to do with the bad guys' mules. It went back and forth for fifteen minutes until I couldn't take it anymore.

"Stop! Enough! You guys are sitting on a multi-million-dollar dope deal. We're running out of time, and we're arguing over a bunch of *fucking* mules. You can tell the prosecutor the Americans let the mules go. Or, Paul and I will shoot the mules, and you can take pictures. We can then build a big-ass bon fire and barbecue them. I don't care, but this issue is over because we've got to go."

Paul found the whole exchange funny as hell, but elected not to weigh in. It was too entertaining. Only in Afghanistan. However, the positive thing to note about their concern for the mules, was that they were trying to follow Afghan law. I couldn't fault them for it but there were higher priorities at the present time. If we didn't get going, there would be no mules to argue over.

● ● ●

The colonel, Sammy, and I met with the general and explained the case. Time was of the essence, so the general picked up the phone and tasked the counter-narcotics colonel with immediately assembling a raid team. We thanked him and beat feet to the Counter-Narcotics Division. After giving him the rundown, the colonel advised he would have a crew ready to do battle in one hour. We would all link up at 9:00 p.m., there at police headquarters.

Our equipment check went smooth at the safe house. Everyone had loaded for bear on this one. We issued Nazar an AK-47 to carry, which made his night. Sammy had brought his AK as usual.

...Let me address this issue for a minute. American rules and regulations prevent interpreters from carrying weapons. That's one of the most absurd things I have ever heard. Some of them have fired more rounds than any U.S. soldier ever thought about. They go in harm's way just like any other soldier. The U.S. government and the corporate attorneys expect them to go unarmed? Sure they do, because they're not the ones running operations in the middle of the night. It's real easy to make policy like that from the comfort of your cubicle in the Pentagon, while sipping on Starbucks. In our unit, commonsense always prevailed. I made sure of it. Therefore, my interpreters rolled armed. Life and limb should trump policy—every time.

I grabbed my extra grenades from the supply room. I usually carried only one, but it wasn't for the bad guys. It was for my own plan Z, in the event things ever got that bad. Tonight, the extras might come in handy.

As we pulled into police headquarters, a familiar scene was taking place. Twenty young ANP officers and several counter-narcotics officers were preparing to do battle. After a quick coordination meeting, we were rolling in masse toward Torkham.

At around 9:45 p.m., we linked up with the two informants who had been sent by The Professor. They led the way and we took a right turn off the main road. After traveling on the paved road for about ten to fifteen minutes, we turned left onto a dirt road. We continued to travel in a vacuum of dust for what felt like forever.

"Are we in Pakistan, yet?" someone asked.

"Seems like we're in India by now."

We began to slow down, and we realized there was a small camp of tents in the darkness, to the left side of the roadway. I say roadway, but there really wasn't a road where we were. We had been blazing our own trail. The camp was inhabited by a handful of men, who were *supposedly* doing a project in the area for some NGO. The camp looked more like a low-key observation post to me. Regardless, the vehicles would stage at this small encampment and several officers would be left behind as security. Once we hit the bad guys, the vehicles would come forward to a rally point in order to rapidly load the dope.

We got out and all took the time to piss. Afghans piss while squatting, so the folks in the camp immediately pegged Paul and I as gringos. I had my M-4 slung and wasn't going to try and squat down. Besides, when I did squat to maintain my cover, I would end up pissing all over my man-jams.

As the members of the team retrieved weapons and ammunition from the vehicles, Sammy and I met with the colonel for one last get-together before we departed friendly lines.

"Sir, we've got a problem," said Sammy, after a whispered chat with the colonel.

"What's the problem?"

"It's the two snitches. They want to talk to you."

"Why can't they talk to the colonel?"

"They already did, but he told them they had to talk to you."

"OK, tell the two gentlemen to come on over and let's see what the issue is. But, we're running out of time, my man."

I began to run through my head all the things they were going to want. Problems with informants always pop up at the last minute during dope deals. Back in the States, its either they're nervous, they're scared, they're hungry, they're thirsty, they've got to call their girlfriend, etc.

"Sir, you're not going to like this, but they want three bags of the hash for themselves."

"What? The Professor knows I got him paid the last time, and I'll get him paid this time," I said.

I did get him some money from the military on the first case, but only after two months of headaches, paperwork, politics, and bureaucracy.

In reality, I knew it would be a long shot getting any more money out of the gringos, especially since I hadn't informed anyone about the op. It would hurt egos and therefore, no checks would be written on my behalf.

"I'm just telling you what they said."

"I know, my friend. Let me ponder this shit for a minute."

I looked at the two informants. While they had an uncertain look about them, deep down they knew they had me—check, and mate. What were my options? There we were, in the middle of nowhere, in the middle of the night, with about thirty troops who had been commandeered for this mission. There was a lot of manpower invested in this deal. The informants were smart enough to recognize it and knew it would not make me look good to return empty-handed.

I had the option of putting guns to their heads and telling them to march toward the trail in question, but there were too many witnesses. I ruled that out. The other option was to tell them to screw themselves, and call off the operation. That decision would allow $3,000,000 U.S. dollars to reach the pockets of Mangal Bagh. The third, and only real option I had, was to deal.

In the States, giving dope to informants to facilitate a drug operation would get me indicted. But, this wasn't America. This was Afghanistan, in the middle of the night, close to the Pakistan border.

Second, third, and fourth order effects of this load getting through meant that young American soldiers may die. There would be thirty-eight to forty bags of hash on those mules. It made sense to give up three, to seize thirty-five. It would also solve my problem of compensating these gentlemen due to the lack of cooperation from the military. Again, commonsense prevailed over policy and bureaucracy.

"*Fuck it.* Tell them it's a deal. They've got my word."

Sammy relayed the message and they responded.

"Sir, they want one other thing."

"They're not getting it, if it's anything complicated. What is it?"

"They said that a couple of the mules in the train belong to them. They want their mules back."

"I'm tired of arguing about these damn mules. Tell them that when it gets taken down, I'm going to slap those mules on the ass and they'll head for the barn. After that, it's not our problem. That's the best I can do. Now tell them to get to steppin'. Negotiations are over."

With that last translation, the two informants knew not to ask for anything else. They set off in the lead, with the column of men and firepower behind them. As dark as it was, on this occasion, we didn't bother dressing them in ANP uniforms.

The stars got brighter the farther we walked. After about an hour of patrolling in silence, we had reached the T intersection. It was in the middle of nowhere, but the faint sounds of dogs barking could be heard in the distance. There was a small mountain in the direction from which the bad guys would be traveling. In the night, its black silhouette made looking in that direction, down the trail, very difficult. It was like staring into nothing. The intersection was on the high ground, compared to the trail from where the bad guys would be coming. As the mule train would approach the intersection (traveling uphill), the right and left trails branched out onto flat terrain.

I had tried to weigh-in on where the men should be positioned and how the ambush should be laid out. But, the Afghans seemed to know what they were doing. My job wasn't to micromanage. It was to enable. It was their country, so we sat back while they positioned men, machine guns, and RPGs. The majority of the force was placed along the trail of approach, in a line ambush on the left side (bad guys' left). Paul, Nazar, Sammy, and I found a nice bomb crater above the T. The position allowed us to look straight down the avenue of approach. From our position, the majority of the force was to our right. There were a few troops about a hundred meters to our left as well. It was a sloppy L-shaped ambush, but it would work just fine.

Everyone was in place at around 11:00 p.m. Sammy and I covered off down the trail and like everyone else, started playing the waiting game. If you've ever sat on surveillance or lay in ambush, you know that you're real excited when you first get set up. You scan everything. After time begins to pass, it is human nature to get bored—I don't care how dedicated of a soldier you are. You start to relax and notice other things. You start to daydream. In some cases, you start to dream.

ZZZzzzzzzzzzz, began to echo within the bomb crater.

Sammy and I looked hard left. Paul and Nazar were laid back against the front of the crater, and were fast asleep. It would have made too much noise to get up and awaken them because they were out of arm's reach. We just started throwing small rocks. We only succeeded in stopping one snorer and jump-starting the other. Bigger rocks were thrown until the snoring ceased, but they were still sleeping like babies. Their shiny, white man-jams up against the black dirt made me laugh. I had suggested they wear poor-man's clothes like me, but they insisted on getting tailor-made man-jams. They would be ruined after this mission. I felt bad, sort of, but it had proved my point. Now, instead of looking like bankers, they looked like dirty bankers.

"Screw it, let them sleep," I told Sammy. "We'll wake them up in a bit. We may be out here all night, anyway. If Paul is well rested, he can drive us back in the morning if the deal doesn't go."

● ● ●

A cool breeze was blowing from the direction of the trail, which gave the night an extra chill. It was beautiful staring up at the stars on such a clear night. Other than the sound of the breeze and a couple of dogs barking in the distance, no sounds could be heard. That was, until a Predator drone came flying over, heading toward Pakistan. The familiar buzz of the engine left no doubt as to its identity. Sammy and I watched as it approached the border area. They say you can't see it, but you can. On this night, you could see its silhouette pinging in and out against the backdrop of bright stars. I breathed a sigh of relief when it didn't stop to investigate us. I didn't have the time to properly notify all the gringos because time was not on our side. I suspected they would probably engage us if they were to spot thirty guys lying in ambush, in the middle of the night. It never occurs to the Americans that the Afghans don't answer to them. They're not required to notify the U.S. military of their actions. It's their country. But the Americans have a different way of thinking. Remember, it's *their* battlespace and thus, *they* own the ground. It's pretty arrogant and imperialistic, if you think about it for a minute.

Therefore, they might just assume we were bad guys and call fire down on us. It happened all the time to private security contractors dressed like locals, throughout the country. Both Sammy and I gasped as the Predator made a U-turn and started coming back toward us. *Had they spotted us?* If so, there was nothing we could do other than pop infrared (IR) chem-lights and turn on our infrared strobes. I popped the 9-volt battery onto my IR strobe, known as a *Firefly*, and dug out three extras that I carried. I popped the batteries on those as well. Whether they worked or not, I'll never know. I didn't have night vision goggles with me to verify they were functioning. None of my agents had night vision, so I didn't carry mine. They didn't have body armor, either. Therefore, Paul and I refused to wear ours.

The Predator continued toward Jalalabad for a few minutes, made another U-turn, and then headed back toward Pakistan. As it went out of sight and sound, Sammy and I both grinned at one another. Apparently, they weren't focused on us at all. Or maybe they did see us, and just had better things to do that night.

However, on a previous occasion several months before, we had been focused on. An observation helicopter had begun to circle Sammy, Rene, and me, as we were about to eat lunch in an orchard. They were apparently looking for a point of origin (POO) site of a 107mm rocket attack, and we were the only unlucky bastards in the area. We briefly panicked, because we thought we were about to get engaged. Our hasty response was to put all the food we had, out on the hood of the car, to blatantly indicate a picnic was imminent. The helicopter eventually lost interest and quit circling. I'm not going to lie, that incident scared the hell out of me. As a result, I consulted with a helicopter pilot about what actions we should take to avoid being fired on, should we inadvertently get targeted while we were on a mission. I put together a safety class for my team. We practiced our procedures in the event we were engaged, to let the pilots know we were friendly. I won't divulge our tactics, but the last ditch effort is worth mentioning and is far from being secret. Partly to lighten the mood and partly serious, I had taught all of my agents how to do the dance to the Village People's song, the YMCA. If all of our safety procedures failed, and we were still getting strafed with .50 caliber machine gun fire from above, we would all start doing the YMCA. I'm not kidding. If an American pilot continued to shoot at a dozen cats in the middle of Afghanistan, kicking it 70's style while under fire, he deserved to get disciplined for a lack of commonsense, after our funeral. Besides, once he saw that spectacle, he probably wouldn't be able to shoot straight due to laughter. Thankfully, we never had to resort to the last safety procedure.

As I lay there in that crater, stargazing, I realized it was one of the most peaceful times in my life. I was thinking to myself, *There's no adventure that will be able to top this one, regardless of what's about to transpire.* It scared me to think I would one day have to return to the United States, and try to lead a normal life again.

● ● ●

At around 1:15 a.m., Sammy's radio crackled. He had it low in the hole and covered, so it couldn't be heard by anyone but him.

"The guards are coming. They're carrying AK's. Get ready," Sammy whispered, as he readied his rifle.

We hurriedly woke Paul and Nazar by throwing a dirt clod.

"Get ready, dog. They're coming," I whispered.

The officers at the front of the line had spotted the guards coming up the trail. They could see the outlines of rifles being carried. They put the word out and went back to noise discipline.

...Let me back up a minute. During the operational planning meeting, we discussed what to do about the armed escorts running block. If we engaged them, the mule train would turn and run. As it was in the middle of the night, it would not be smart to try and pursue them through unfamiliar terrain. They knew the ground and we didn't. The ultimate objective was to interdict the load of dope, thus taking the proceeds out of the bad guys' pockets. Simplicity. Take the load. The guards were nobodies who were probably paid the equivalent of fifty U.S. dollars to escort the load. The guys driving the mules were nobodies as well. Even if we grabbed them, we were just grabbing poor people trying to make a living. We eventually made the decision to let the escorts pass right through the ambush, and keep on going. It was the only way to be sure to hit the load. Once it went down, we would obviously be alert that the guards might come back and fight. Hopefully, once they discovered the size of our force, they would deduce that fifty bucks wasn't worth dying over. To be completely honest, I didn't think the plan would actually work. For it to be successful, it meant that the bad guys running point had to walk right through the middle of our ambush—right through the middle of around thirty men, lying in the dirt. It meant that noise discipline had to be maintained, without fail. If one person coughed, sneezed, farted, or clanked his rifle as they walked through, it would instigate a firefight. We would have easily taken the guards, but the mules would be nowhere in sight. While I was trying to remain optimistic, I just wasn't sure we could be that quiet.

Several minutes later, we saw the outline of the guards coming up the trail. They veered left (to my right) and continued walking at the same pace. As they passed within meters of our position, I could see the outline of AK-47s being carried at the ready. They slowly faded out of sight. So far, the plan was working. The gentlemen had no idea they had just passed through the kill zone of a thirty-man ambush.

Ten minutes later, Sammy's radio again crackled.

"They're coming!" Sammy whispered.

He didn't have to tell me, because I smelled them. The breeze that was blowing directly toward our position carried the smell of sweaty mules with a tinge of top-quality hash. It was at that moment, I knew we had our quarry in our sights. Now, it would be up to the bad guys as to how the scene played out.

DRAYYYYYSH!!!! DRAYYYYYSH!!!! The commander began to yell commands, once the mule train was deep in the kill zone. The word *draysh* is a term that only the police or the military use, to tell someone to halt. If one hears this, they would automatically assume it's government troops and not thieves.

Sounds of running feet pounding dirt, filled the air. Commotion ensued. The sound of slings clanking against weapons was heard. Dust began to bellow. Controlled chaos was intertwined with yelling in Pashtu. We could feel the thump of hoof beats from our fighting position.

The bad guys elected to flee back down the trail as planned—they didn't care to stay and fight. They may not have known it, but they made the smart decision. In accordance with the operational plan, nobody was baited into pursuing them into the darkness. The line was held.

In a haze of dust during the commotion, the now free mule train topped the hill and popped out right in front of us. As if on auto-pilot, they made a right turn and kept on trucking toward Pakistan. They had obviously made the trip many times. We could see they were loaded down with rice bags, full of dope.

The four of us wanted to take off after the mules. It was instinct.

"Stay down! Everybody stay put and let it play out! If one person starts shooting, everybody will start shooting."

We stayed put in our makeshift foxhole because it was the smart thing to do. While the mules were escaping with our dope, it wasn't yet worth moving from the perfect fighting position. If the bullets started flying, we were golden. Once we saw a few of the Afghans getting up from their positions, it was time to wrangle some mules. The four of us took off at a sprint down the trail to our left, with a couple of ANP officers. We finally caught up with the lead mule and got the train stopped. The mules were overloaded to begin with, and were sweating and out of breath. They were decorated ornately, just like the donkeys you see in old Mujahideen footage, when they were smuggling Stingers in from Pakistan. The whole scene was nostalgic.

We quickly drove the mules toward the rally point. The convoy commander radioed for the vehicles to come forward, and their headlights cut open the darkness as they approached. The colorful webbing that held the bags onto the mules was hastily cut with knives. *THUNK!* It was the sound of several hundred pounds of hash hitting the dirt after being cut off of a mule's back. While the team cut the webbing, a young officer began to yell out into the darkness. He and several others took a knee and pointed their rifles in the same direction. Paul and I turned to see the silhouette of several figures converging on our location. We immediately took cover behind the only thing we had in front of us—four bales of Taliban hash. The bags were as hard as set concrete and would have stopped a fifty caliber round. Everyone feared these guys were the security team for the load. A yelling match ensued, but the group ultimately elected to disappear into the darkness. After that scare, the pace quickened and soon we had thirty-five bags of hash loaded into the ANP Ford Rangers.

During the confusion, three of the bags had been quietly and covertly concealed in the back of my Forerunner for transfer to the informants. A deal is a deal, and your personal guarantee in Afghanistan cannot be broken. Among the locals, there is no need for a written contract. If you give a person your personal guarantee, it is signed in blood. If a personal guarantee is broken, people die. Family members die. It is the way.

●●●

While some of my actions may not sit well with U.S. attorneys and uptight FBI agents who are driven by policy, all I can say is, *tough shit*. You weren't there. I had made a commitment to myself, that if I had the opportunity to trade my life to save that of a young American soldier, I would gladly do so. I was coming up on forty, and had already lived a good life, anyway. It was the right thing to do and the right mindset to have. Therefore, it was my mission to protect those young kids in any way I could, even if it meant violating a few policies. Since I was prepared to give my life, I certainly wasn't concerned with lesser ramifications I could have faced, as a result of a decision made under duress.

Take a minute to put yourself in my shoes at that point in time. Would you give a Taliban commander three million U.S. dollars, who will inevitably use some of that money to kill young American soldiers? Or, would you give ten percent of that money to a couple of farmers, who will use the money to feed their families? You have the luxury of time to come to a conclusion and formulate a plan. I wasn't afforded that luxury. *Make a decision.*

CHAPTER 12

Typical Days

I told Sammy early on that we weren't doing any work on Thursday afternoons. Since the team was off on Fridays, Thursday would serve as our Saturday night. On Thursday mornings, we would load up the Forerunner with bags of rice, beans, and flour from our shipping container, and make a run to the homeless camp I had adopted. Good Samaritans back home were sending tubs of children's clothing as well. The back bumper of the Forerunner would almost drag the ground by the time we finished loading it. I had chosen a small homeless camp that was inside the city, near the river. It was actually a refugee camp, because most of the folks were from Pakistan and had fled the violence there. I chose it because it backed up to a swamp which served as a blocking area. We had two ways to ingress/egress in case things got out of hand. There were only about thirty different families there, which made it manageable for our resources at hand. I was determined to make a difference in their lives. Dealing with bad guys was my job, but helping my people in the refugee camp was what I really enjoyed doing.

The first time we rolled into this particular camp was quite an experience. Obviously, they had no idea Sammy and I would just show up and start handing out humanity. Our first load was comprised of children's clothing. When I stopped the vehicle, no one reacted because they didn't know what was going on. Sammy and I had rehearsed what we were going to do, so he asked everyone to sit down and then we started the distribution. While it was a bit chaotic, we handed out the clothing and were able to take some pictures of the process. We rolled out of there feeling we were indeed, doing good will.

The next week, I decided to start getting rid of the food in the shipping container. Better my homeless folks eat it, than the hoard of rats. Everything else could wait to be distributed, after the food was depleted. Besides, food was the immediate need. We rolled into the camp and found ourselves surrounded. We got out of the Forerunner, closed the doors and locked them. The rehearsal called for Sammy to establish order before any distribution began, but on this occasion, there would be no order. I would quickly discover that charity work can be more dangerous than chasing bomb makers. Chaos erupted. A brawl ensued over the pecking order to receive the bags of rice that could be seen in the hatchback. Things were immediately out of control to where we had to resort to escape and evasion. We barely made it back inside the vehicle without being mauled ourselves.

"Shit! That wasn't supposed to go down like that."

"Yes, that was crazy. They cannot act like that."

"Let's go to Darunta and drink a few. We've got to figure this out, because we're too smart for that to happen again."

With that, Sammy and I developed course of action (COA) number two, for combat charity operations. The first thing we would do was to deploy back to the camp, without any food or clothing to hand out. We would identify a director of security to maintain order during the distributions. We would lay down the law to the director that if anyone got out of line, we would leave and distribute nothing. We were there to help them, not to cause fights. It would be to everyone's benefit to remain civil while we were there.

A few days later, we parked the Forerunner about a half-mile down the road at a restaurant, and walked in. No one paid any attention to us until we were already in the middle of the camp. The good folks started scanning the area for the magical Forerunner they assumed would contain provisions.

After a quick search and brief interview process, we had found our man. He was a middle-aged guy of average intelligence. He only had one leg, and coincidentally, so did his brother. They only needed one pair of crutches between the two of them. We sat down outside his tent and laid out our requirements. The first thing we needed him to do was compile a list of names of all the people who lived in the camp. I wanted to know who belonged to which family, and who the head of the household was. Once he had it complete, we would return and pick it up. From his master list, I would prepare distribution cards that had numbers on them. Upon our arrival to do a distribution, the good director should promptly have one person from each family line up in numerical order, with their card in hand. Only people who possessed cards would receive assistance. It had to be that way or else word would get around and we would be overwhelmed. I had it in mind to only make thirty cards.

When we went back to check on the progress of our list, the one-legged director had it complete. Luckily, there were only twenty-six families on the list. That left me with four extra cards to hand out to various families we had run into during our travels. I took the list and later transposed the names onto the recipient cards. Now we would have some semblance of order.

The cards were issued out on a Wednesday, and Sammy gave our director a pep talk. *We had better see some organization when we roll in tomorrow,* he was told. He insisted he would have law and order upon our arrival. Just bring some food. The clothing and blankets would be great, but what they really needed was something to eat.

"Comandante, we are counting on you. Don't let us down. You keep the peace, and we'll bring the food."

The next day, we rolled in slow, with a Forerunner loaded to the gills with rice. As promised, our security director had everyone in line with their cards in hand. We opened the hatch and began an orderly distribution. The only problems that would arise would be from the curious children trying to get too close and be a part of the action. The adults did a pretty good job of keeping them back, and we were able to get rid of all the rice in an equitable fashion. I would learn that you never give one person more than another. If there was rice left over, we took it somewhere else to maintain the balance. Inevitably, there were new people who had shown up to see if they could get assistance. We couldn't help them because we just didn't have the resources. Had we tried, we would have turned the small camp into a bustling metropolis of people looking for aid.

This began our weekly routine of doing charity runs to the homeless camp. Imagine entire families living in makeshift tents made of sheets that had been sewn together. Elderly women, newborn babies, and small children were sleeping on the ground. The toilet was the swamp out back. Mosquitoes were rampant. They were living in absolutely horrible conditions, and the weather was cold at night. Wherever I lay my head, my mind would continuously wander to the camp. My stomach would turn when I thought of those poor children freezing in the elements. It just made me work harder to get them basic provisions. We would end up taking them tarps, small cooking stoves, blankets, clothing, rice, flour, shovels, shoes, tea, and other items from the shipping container. In addition, we had boxes of children's clothes that were sent from the good people of America.

...Religion truly divides us from humanity. My wife (a devout Christian) had been in charge of collecting goods and shipping them to me. She told me that her Sunday school class wanted to make the children some blankets. I thought it was a great idea, until she said they wanted to embroider "Jesus Loves You" on them. I didn't want to create a story for some bored journalist, so I told her I didn't think it was a good idea. It could, and would have been interpreted that I was attempting to spread Christianity. Subsequently, they lost interest because of the issue and never quilted the blankets. Looking back, I should have told them to put anything they wanted on them. A freezing child doesn't care how he gets warm. I should have said to hell with religion and political correctness. Just help us help these children and quilt the damn blankets. While you're at it, go ahead and embroider a big-ass picture of Jesus nailed to the cross, next to the "Jesus Loves You" part. The heavy embroidery should up the insulation factor and make for a warmer blanket. Humanity should have trumped their desire to spread Christianity.

We were basically supporting thirty families. After a week's worth of hunting bomb makers and running the gauntlet, my refuge was the time spent with the citizens of my homeless camp. Sammy and I are just a couple of rednecks, but in their eyes we had been sent by Allah. They told us they prayed every day for our safety. They probably still do.

● ● ●

After our homeless camp run, we would roll to Darunta and relax by the river. Actually, we would get drunk as hell by the river. Afterward, we would meet up with our friends from various government agencies, intelligence agencies, police agencies, village elders, politicians, etc., and have dinner. This is where I would learn what was really going on in the world. No intelligence report could beat what I learned during those all-nighters. They couldn't even come close. I would even be privy to opinions about their past week's interactions with the Americans and their silliness. Except for the absence of women, those were some good times I'll never forget. There's nothing like getting drunk with the village elder, who the gringos just had a key leadership engagement (KLE) with. (A KLE is a complicated term the military uses to describe a meeting with the locals.) After listening to the ridiculous information operations

(IO) message of the week, the village elder needed to drink a few (at least that's what he said). They knew the Americans were putting on a dog and pony show at these meetings, but so were they. They planned out what they were going to say, just as the Americans did. It was mostly smoke and mirrors, to match smoke and mirrors—formality versus formality. When the gringos left, they would have a real meeting in Darunta and talk about the issues over dinner.

"How many times are these fools going to ask me if I've seen any Taliban?" asked one village elder. "The only thing the Americans care about is finding someone to shoot at."

I told him that I didn't have an answer for him. In reality, I just *chose* not to answer him.

● ● ●

We were relaxing one night at the safe house after an uneventful day. The two duty agents, Mr. G, the young soldier, two other agents, and I were all watching an Indian movie on the television. Agent M's cell phone rang and he immediately sat up from his relaxed position. His voice got louder in conversation with the caller. I looked to Mr. G, who said we had to roll, immediately. As everyone was grabbing rifles and gear, Mr. G gave me the basics. A snitch had eyes on two assholes emplacing a bomb in a culvert on Kunar Road. The location provided was less than ten minutes away from the safe house.

Kunar Road runs north into Kunar Province from Jalalabad, and is a major supply route for the Americans. It was obvious who the intended target would be, as American convoys continually got hit on the route.

Within minutes, two Forerunners were speeding out the gate with the six of us. We had to leave the young soldier by himself to guard the safe house. Agent M was back on the phone with the snitch, who advised there were now three assholes working on the bomb. They were armed with AK-47s. As we neared the area, we killed the headlights and slowly crept toward the culvert. It would end up being easy to locate, because it was about one hundred meters north of a small shop we were all familiar with. Through the darkness, everyone scanned for movement.

As we passed the dimly lit shop, the three bad guys apparently spotted us. They jumped up from the culvert on the left side of the road and ran like hell. They ran north, toward a series of qalats. We bailed out and took up prone positions as the three disappeared around the back of the first qalat. We expected them to stop and fight but there was nothing but silence in the breeze. They probably assumed we had more people with us than we actually did. Either way, it wasn't in our best interest to pursue the gentlemen under the circumstances.

Agent S bound up to the culvert and confirmed there was a bomb inside. The command wire wasn't yet attached and was lying in a tangled mess next to the culvert. The tangled wire had apparently slowed down their progress, long enough for us to get there and interdict their little operation. By now, my team knew the drill and took up defensive positions while I got on the cell phone. I called back to Task Force Paladin and gave them the lowdown. Since we were only ten minutes from town, I requested that Task Force Gryphon send an EOD team as soon as possible, to deal with the bomb. At that particular time we had the scene secured, but obviously couldn't predict if the bad guys would come back in force. As I made this request, I laughed to myself because I knew what the outcome would be. But due to politics, I had to go through the hassle of requesting U.S. EOD assets first. Then, once Task Force Gryphon refused to send anyone, I could contact Ralph. As I waited on the official refusal that was inevitable, I put in the call. He answered on the second ring with a cough. He had obviously been asleep.

"Ralph speaking."

"Ralph, this is Mark. We're out on a bomb in a culvert. The typical political drill, so I'm waiting on Gryphon to refuse to roll. If you could get your guys together, I'm sure I'll be calling you back as soon as I get the official *we ain't coming.*"

"No problem. We'll be there in five minutes once you give us the green light," Ralph said with enthusiasm.

"As usual, thanks my man. Sorry to get you out of bed."

"No problem. I'll have my people waiting in the trucks."

Now that's service. That's a can-do attitude coupled with an excellent work ethic. It pissed me off as I sat there and waited on the inevitable response from Gryphon, while we held a loose perimeter around the culvert. But, I had to wait. As long as Gryphon refused to roll, I could defend the fact that Ralph and his team were my only option. If I called Ralph without waiting on the refusal, I would catch hell from on up the chain. At least five other entities would have something to complain about as well. Thirty minutes later, the call came.

"Yeah, they're not going to roll because it's dark and it might be an ambush. They advised to hold the perimeter until around ten tomorrow morning, and they'll send a team."

It wasn't even worth getting mad at the messenger. He was just relaying the information provided.

"That's fine, my friend. That's the word I've been waiting on. But call the TOC back, and relay some information since they now have our grid. Tell them there are plain-clothes police officers on the ground in defensive positions. If they send a Predator over this way, advise them there are friendly personnel on the ground in man-jams. Therefore, do not engage any targets at this location. I say again, do not engage any targets at this location."

As usual, it was my only real fear—getting attacked by American aircraft. The guys planting the bomb we could handle. A Hellfire missile we could not.

...What pissed me off, *was that the bomb was intended to hit an American convoy, probably a Task Force Gryphon convoy. We prevented one of their vehicles from getting hit, and the way they repaid us was to refuse to roll. They (the leadership) were too afraid to allow their soldiers to roll in the dark, but found it acceptable for me and my Afghans to hold the ground until daylight. It is small incidents like this, that let the Afghans know, the Americans really don't care about them. How could I defend my fellow Americans? They had armored vehicles, machine guns, twenty soldiers, and air support, but were afraid to roll. We rolled with six people armed with AK-47s, wearing man-jams, and driving old Toyota Forerunners. We were still alive. Think about that for a minute, from the Afghan perspective.*

ANP officers in a Ford Ranger made it to the scene to assist with security. About eight minutes after I called Ralph, his convoy of trucks were on the scene as well. They disarmed the bomb and searched the rest of the culvert. I took some pictures for the report and we were soon on our way back to the safe house. Mission accomplished. We didn't catch the three bad guys, but we had prevented an MRAP from getting blown up the next morning.

The next day, a round of e-mails circulated in reference to our operation. They were not congratulatory if you hadn't already guessed, but for some odd reason I didn't take too much heat. Maybe the folks on their e-mail distribution list had grown tired of listening to them complain. Whatever the reason, and at that stage in my tour, I really didn't care anyway.

I congratulated the men on a job well done. On a moment's notice, they had prevented a bombing. They had potentially saved a young kid from getting injured, or even worse, getting killed. It was merely police work in action. They were finally doing exactly what they were being paid to do.

A few days later, we had a similar incident. However, it was daylight when the snitch called. He had seen a couple of men digging a sandbag into the side of the roadway. We hauled ass to the location, but the bad guys were long gone. We located a sandbag that was half-buried, just as the snitch had indicated. I had my agents radio for uniformed officers to come help with blocking the roadway, while I put in the call.

On this occasion, I didn't get the rapid refusal because it was during the daytime. I got the word to hold the perimeter because Task Force Gryphon would be sending a team out. *Well, ain't that a change.* About an hour and a half later, the familiar buzz of a Predator was heard overhead. Everyone began looking around to see if they could spot it, but couldn't. We just heard it.

After about thirty minutes of buzzing, I received a call asking me to confirm my location. I read back the same ten-digit grid I had initially provided. The response was that the Predator couldn't locate us, and they were requesting a better location. *Huh?* How do you give a better location than a ten-digit grid? Just fly to the grid and look for the police trucks with their blue lights flashing. They'll be the ones blocking the road. I thought they were joking with me, but after another twenty minutes of buzzing, I began to believe they really couldn't find us.

Since I had known from the start that we would be playing the waiting game, I had sent some folks on a lunch run. Once again, the funding came from the Bank of Mark. We had plenty of security in place and it was a nice day, so we turned the game of waiting into a picnic as well. It made for a good opportunity to break bread with the young ANP officers, to further our relationship with them.

I'm not going to take up too much time on this incident, because the end result was that U.S. assets did arrive to assist us— six hours later. *Six hours.* However, we were only about fifteen minutes from the FOB the convoy rolled out of. Could there be any good reason for it to take that long? The EOD crew had been ready to roll the entire time.

Once the young EOD technician arrived, he made quick work of the bomb in the sandbag. The obvious outcome, as he's the best in the world. He made me proud to be an American, as I watched him work from a distance. Within thirty minutes, we were taking pictures of what was left of our quarry.

This incident was just another example of too much command and control from people above the level of ground truth. It took six hours to gain permission to travel fifteen minutes. I don't think the Predator ever did find us. They must have given up after five hours of searching and decided to send the convoy anyway.

● ● ●

We had been working with an informant for a few weeks, on a cache that was reportedly buried at the top of a mountain. It turned out to be a politically sensitive case because the *owner* of the cache was a member of parliament. Everyone involved with the case was scared that the gangster, turned politician, would have everyone involved killed. In Afghanistan, that was a legitimate concern. I finally convinced the leadership involved that we couldn't turn and run from a case, just because a gangster owned the land the cache sat on. They had to be the bigger gangsters and raid the cache. There was no other option. To not conduct the operation was to concede that the government was ineffective, and not capable of enforcing the law.

Once everything looked solid, we rounded up our backup from the ANP and got Ralph's team on board. After a briefing in Jalalabad, we headed west out of town. As we were nearing Darunta (the small retreat on the Kabul River), Agent D asked Sammy a question in Pashtu. Sammy laughed.

"What's the deal?" I asked.

"Sir, Agent D wants to know if you're going to buy the beer on the way to the mission."

"What? I'm not going to buy beer on the way to the mission. I need everyone's head in the game. Tell him if the cache is as big as the snitch says it is, I'll buy everyone beer on the way back."

"OK, he says that sounds good. No problem."

An exchange in Pashtu ensued as the convoy pulled over next to the shops near Darunta. Several of the officers got out to buy a Pepsi, or to get a snack before we rolled. Agent D got out and disappeared into the mix, as Sammy and I chatted. A few minutes later, he returned with a trash bag full of items. I immediately heard the distinct sound of the top being popped on a warm Heineken. I turned around as Agent D tilted the can up and gulped it down. He finished with a big-ass smile on his face, as a bit of Heineken dribbled down the corner of his mouth.

"Hey! I said I was buying beer on the way back from the mission!"

Laughter ensued, and Sammy advised, "He said no problem, sir. You will still buy the beer on the way back from the mission. He bought the beer for the way."

I had no choice but to laugh with them. How do you argue with that? The answer is that you can't and you don't. When in Rome, sometimes you have to do as the Romans.

"Alright damn it. Then pass one my way."

With that, we began the two-hour drive to the objective, while sipping on warm Heineken and listening to rock and roll on my boom box. Creedence Clearwater Revival was the music of the day. We turned off of the pavement and headed out through the valley floor, on what you could barely call a road. An hour later, we were at the link up site. There, we would wait for our uniformed ANP brethren who would serve as our muscle. While we waited, we took the time to test fire our weapons and throw a few grenades for practice.

The ANP officers arrived and we conducted an operational briefing. After radio and weapons checks, a ten-vehicle convoy headed toward the base of the mountain. After reaching the base, it was immediately apparent it would not be easy traversing the rocky goat path to the top. The Forerunners and Ford Rangers barely had the turning radius to negotiate the sharp turns at every corner. I had to keep the four-wheel drive engaged during the entire ascent. Looking back at the vehicles straining to keep up, I pondered why the U.S. military was riding in MRAPs in this region of the country. The only way they could get to where we were, would be by helicopter.

The view was absolutely amazing as we continued to climb. We passed a few small villages that were really just clusters of three or four qalats. After struggling up the mountain for over an hour, we literally came to the end of the line. We were the first vehicle, so I stopped about one hundred meters shy of the large rocks and boulders signaling the end. Luckily, there was a small indentation to the right that could serve as a turn-around spot. Once I see-sawed the Forerunner back and forth to where I was turned around and facing out, I put it in reverse and backed down the roadway the remaining one hundred meters. It took about thirty minutes to get all of the vehicles to where they were facing back down the mountain, for a quick getaway. Security was established by placing two machine-gun positions on the ridgeline, overlooking the vehicles. These guardian angels would also be able to cover us during our initial approach to a nearby village.

Everyone geared up and prepared for the patrol to the cache site. The team consisted of twelve ANP officers, eight of my agents, and ten of Ralph's people. Collectively, we were armed with three PKM machine guns, one RPG-7 launcher, six RPG-7 rounds, twenty rifles, a backpack full of frags (M-67 fragmentation grenades), ten shovels, five pick axes, and two metal detectors.

I made the informant lead the way. He was disguised in an ANP uniform and had been *issued* an AK-47, as usual. To the villagers, he was just another member of the patrol. As I've said before, if he was leading us into an ambush, then so be it. At least he would be the first one killed. Sammy and I were right on his heels, followed by the rest of the patrol. We always led by example, by walking point with the snitch. But, walking point served a dual purpose. Not only were we leading by example, we were reading the informant. If anything was about to break bad, we would know well before it happened, just by reading his body language. It wasn't exactly rocket science but it was smart, considering we were relying on a snitch to lead us.

From the vehicle staging area, we initially descended into a small ravine but quickly began an incline toward a small village. The women and children looked on in curiosity as the men came out to greet us. The village elder approached Sammy and engaged him in conversation as we continued to march. He was so excited he could barely contain himself. Sammy chuckled during their exchange.

"He got excited, because he thought we were here to hold a *shura*. He said that government troops have never been to his village before. I told him we were just passing through, but he still invited all of us to come back and have tea."

I wished that we had time to stay because the old man was really excited and happy to see us. I felt as if we were all celebrities walking through that particular village. There was no sixth sense going off, telling me that we were the enemy or that we were in danger. Simply put, I felt welcome there. The women made no effort to cover their faces and looked upon the parade of men and weapons. These were mountain people who were merely trying to live their lives, absent politics, war, and religion.

Our route took us along a goat trail at the edge of the ravine. As we approached a small stream we had to cross, a man jumped up from out of nowhere on the far side, and took off running. It scared the daylights out of us and we hit the ground. The patrol followed suit and went firm. Sammy and I had a good bead on the guy but he just kept running.

"Look at that son of a bitch go! Damn he's fast," I told Sammy.

After the words left my mouth, I realized I had subliminally referenced the movie *Forrest Gump*. The only difference was that this unknown, barefooted Afghan (he had run out of his sandals), was faster than Forrest. He's probably still running today. We held our ground for a few more minutes just in case and then continued the patrol. We crossed over the stream and began to traverse the incline to our right. About halfway up, the informant stopped and pointed to the ground. I thought the guy was crazy or that maybe he had led us into an ambush, but he insisted the cache was where we were standing.

● ● ●

We set security in place and Ralph's people commenced to digging. After ten shovels worked for over an hour, they hit the cache. The first piece of ordnance to come out was still in its plastic case. Ralph opened the plastic case to reveal a Chinese-made 82mm recoilless rifle round. I felt better about the situation at that point, because we were in the right place. As Ralph's people kept removing ordnance, we discussed how we were going to get it out of there. I initially thought the cache site would only be a few hundred meters from the vehicle staging area. That meant we could have carried the ordnance back to the vehicles, but instead, we were at least three clicks away though rough terrain. As they pulled out the hundredth piece, it was apparent we would have to go to plan B.

"Well Ralph, we can go back down to the village and commission every available donkey to ferry this stuff back to the vics, or you guys can blow it in place. What do you think?"

"Let's see how many more pieces there are, and then we'll make the call," Ralph replied.

After they pulled out another hundred rounds, it became apparent that we had too much to ferry back to the vehicles.

"Yeah, looks like we're going to have to blow this in place," said Ralph. "I'll send some of my guys to the village to hire a donkey to ferry the charges back to us."

With that, Ralph's people and a few ANP officers rolled back into the village. They commissioned a young Afghan and his donkey to carry the explosive charges and equipment, back to the cache site. About an hour later, the crew returned with a small, white donkey in tow. The poor donkey was loaded down with crates. Donkeys are the reason the Mujahideen kicked the hell out of the Russians and why the Taliban is kicking the shit out of us (the Americans). A donkey is durable, cheap, can climb mountains, and doesn't break down. A million-dollar MRAP has none of those qualities. The U.S. military should sell one MRAP and use the funds to buy a million dollars-worth of donkeys. Then sit back and evaluate the results.

I decided to call Task Force Paladin and have them pass word to the gringos that we were going to blow the cache. I got the first sergeant on the phone and told him what we had.

Excitedly, he said, "Cool, man! You need me to roll an EOD team out your way?"

The first sergeant was a good man.

"Top, I'd love for you to, but they just can't get here."

That was a true statement. They could drive to the base of the mountain in their MRAPs, but it was too far to the top for them to hump. Definitely too far from them to hump, while loaded down with equipment. Even if they could get airlifted in (which would take two weeks of coordination), there was no suitable landing zone anywhere in sight. They just couldn't get there. The only thing the first sergeant could do for me, was to call Task Force Gryphon and let them know what was going on. It would give them something to gripe about the next day.

...I could have met the soldiers at the base of the mountain and transported them to the top in my Forerunner. However, no U.S. military commander would have given approval for that course of action. The brass in D.C., drinking their Starbucks, would have heart attacks if they knew soldiers were outside the wire with a bunch of Afghans in Toyotas.

The final toll of ordnance that came out of the cache was two hundred sixty-five 82mm recoilless rifle rounds, fourteen PG-9 rounds, and twenty OG-9 rounds. The informant kept insisting we had only scratched the surface of the cache, and that there were two thousand more pieces in the hole. The problem was that we were running out of time. As Ralph's EOD techs began to prepare what we already had for detonation, the sun was quickly going down.

As insistent as the informant was, I came to believe there was more that we were missing. However, ahead of us was a long hump back to the vehicles, in the dark. Afterward, we had a long trek down the mountain, the same way we came. If there were any bad guys in the area, they would have had twelve hours to dig bombs into the road. We couldn't take an alternate route because there was none. After our cache went boom, it would attract even more attention. We would destroy what we had at the moment and live to fight another day.

As the sun went down, Ralph's people completed the prep work for the shot. We ordered the majority of the patrol to head toward the vehicles. They would warn the villagers so they could secure their children and animals. The rest of us would follow after the shot was complete. Sammy, Ralph, one of Ralph's EOD techs, and I took over the machine gun position on the nearby ridgeline. *Fire in the hole! Fire in the hole!* The warning was sounded in English and in Pashtu. Ralph's man hit the trigger as we looked up and out from our cover.

BOOOOOOOMMMMM!!!!! The ground shook, even though we were behind the crest of the ridgeline and a few meters down the back slope. After dirt, small rocks, and debris rained down on us for a few seconds, it was mission accomplished. The four of us quietly beat feet back to the vehicles in the pitch black of the darkness. The convoy lumbered down the mountain without any opposition or conflict.

● ● ●

We took an alternate route once we reached the valley floor. The problem was that no one was familiar with the way, and we ended up getting lost and had to double back. Our route took us through several villages which had recently seen a bit of flooding. During the mud bogging in the dark, one of our vehicles got separated from the group. Sammy got a call on his cell phone and told me we had to go back. We advised the convoy to go on without us, as we turned around to assist. I felt safer being anonymous, anyway. A convoy just attracts attention and I would rather take my chances on my own.

We came upon our missing vehicle in the middle of what appeared to be a small lake. However, it was just a low-lying area that had been flooded. Somewhere in the mix was in fact, the road. Our vehicle was sitting in the middle of the lake, off all four tires. It had bottomed out on a sand bar. They had been trudging through the mud, when all of a sudden, it was as if they had hit a brick wall. A couple of the agents were out in the mud trying to push, but it was no use.

I thought we were screwed and may just have to leave it until the next day. But within minutes, several villagers had come out to help. It was about midnight by that point. I sat there and watched this display of teamwork and unity, from the comfort of my vehicle. (Since I was driving, I stayed in the vehicle with the weapons, and as a side benefit, I stayed out of the mud.) A few minutes later, I heard the sound of a tractor off in the distance. Shortly thereafter, the tractor arrived with three men riding on it. They drove through the mud and backed up to the stranded Forerunner. After getting muddy as hell while attaching a chain, the good Samaritans pulled our vehicle out of its predicament. After a bunch of hugs and hand shaking, the ordeal was over and we were back on our way. Even though everybody was covered with mud (except for me), they were full of laughter about the short adventure.

...It was small incidents like this, which affirmed my opinion of the Afghan people. There was no obligation, benefit, or motivation for all those folks to come out and help a carload of total strangers—especially at midnight, for free. But, it's the way they are. They are just good, simple people.

We finally made it out to a paved road in Kunar Province. We took Kunar Road down to J-Bad and called it a night.

● ● ●

The next day, the whining, drama, and quarterbacking from the gringos began. E-mails were circulating about why I didn't use U.S. EOD assets, and why Task Force Gryphon wasn't notified prior to the operation. I was navigating the political minefield of a successful mission, once again. I wanted to respond to the queries by merely saying, *How about you all just kiss my ass and shut the hell up.*

To stir the pot even more, I decided to *recommend* that the U.S. military conduct an airstrike on the same location, in the event we had not blown the entire cache. This was met by a flurry of resistance, dissent, and opposition. Theories about airstrikes only scattering ordnance were cited. It was too dangerous, etc. A plethora of people sitting in front of their computers with nothing better to do, seized the opportunity to weigh in and offer an opinion. Besides, it would make them appear as if they were actually doing something, if they shuffled some e-mail traffic around to their bosses.

The truth was, I already knew the arguments. I just wanted to bring attention to the fact that there was probably more ordnance in the hole. Had I merely stated I thought there was more, it would have received no attention. By *recommending an airstrike,* it stirred the cubicle bandits and the sergeant safety types, thus shedding light on the subject. They accomplished the mission for me. What's the old saying? *Work smarter, not harder?*

In response to the heat I took over not using a U.S. EOD team, I angrily drafted and sent the following e-mail:

> Gentlemen,
>
> I agree that U.S. EOD is the preferred response to EOD/IED incidents. They are the world's subject matter experts, are highly trained, and are equipped with the best equipment.
>
> With that said, let me break down reality. The battlespace "owners" have the luxury of deciding when and if to send EOD elements. EOD is always willing to roll, but the battlespace "owners" will not allow them to respond. In some of the areas in which we operate, they can't get there anyway.
>
> The ANP have no choice but to respond. Much like cops in the states. When they respond, they do not want to hold a perimeter in Taliban infested areas overnight, or for two days waiting on CF's to arrive. Often the case and I don't blame them.
>
> For example: Two days ago we had a snitch call and advise us that he had eyes on three individuals digging an IED in the road. Five minutes later I notified the Brigade. The next morning, the CI called and wanted to know why no one responded. We dispatched ANP. ANP rolled out, but no CF response. They departed. The CI called back this morning and wanted to know why the bomb was still in the road. We re-dispatched ANP to respond. (ANP has to verify the IED before CF will even think about responding.) ANP got hit with an IED while en route to the IED our snitch reported, and never made it to the scene. End result? One ANP Officer injured, one IED still emplaced, and one perplexed CI. Welcome to my world.
>
> So please advise the chain of command that we would love to have U.S. EOD work with us. It's just not possible here. WRA is a life saver on a daily basis for us, the ANP, and the citizens of Afghanistan. We should be thanking Ralph Ward and his team every day instead of quarter-backing them and their methods. The ANP, ISU, and the personnel from WRA are the true Battle Space Owners, and should be respected as such instead of suppressed.

Sorry to vent, but Ralph and I have endured too much bullshit from people sitting in front of computers and creating Power Points while we were out handling business in harm's way. You cannot calculate the amount of lives that have been saved by ISU working with WRA. But no one seems to understand that.

Another point: most of these caches are buried and require intense manual labor to locate them. Ralph's personnel do a lot of work. Work that U.S. EOD personnel would probably not do, or could not do wearing the required PPE. We don't just show up and blow the cache in place. We have to deploy numerous sophisticated weapon systems first. The first weapon that's used is called a shovel (U.S. made, 1 each). After about an hour of deploying that weapon system with a 15 man element, then you reach the cache. Another weapon that we often use is called a donkey (sometimes brown, sometimes white in color). The donkey is used to carry the charges needed to blow the cache in place because it is too far from the vehicles for a person to carry. Definitely too far for U.S. EOD personnel to carry over mountainous terrain while wearing 100 pounds of gear. Now I'm not sure, but I don't think U.S. EOD MRAPS are equipped with 15 shovels, nor a donkey. Mandatory equipment for our work. UPDATE: I just confirmed with an EOD Tech that they are only equipped with a SINGLE shovel on their MRAP. Not sure what their contingency plan is in case the handle breaks. WRA is equipped with 15 plus.

I can take Ron's advice and start indicating in the reports exactly why U.S. EOD was not used. I could state facts, but I'm sure it will just create problems for us here when the ground truth is published. Politics. Therefore, I will continue to operate in stealth mode as usual in order to save lives and accomplish the mission. The question will continue to be asked....

Here in Nangarhar, I am waiting for the new unit to RIP in, because the one covering Nangarhar is combat ineffective due to poor leadership at the top. You gentlemen be safe.

Respectfully Submitted,
Mark Blackard

CHAPTER 13

Stealing Heroin

Sammy had drummed up a new informant while we were out and about one day. We really didn't know anything about the guy, other than the fact he was from the Bati Kot area. He called a few days later and wanted to meet. We agreed to meet him in town. Sammy, Agent M, and I rolled out to pick him up. He was waiting for us in the bazaar which was extremely busy that day. I parked the Forerunner on a side street while Sammy called him on the cell phone.

"OK, sir. You wait right here and let me go see if this *motherfucker* is straight," said Sammy.

"Be safe, my man. I'll be right here," I replied.

Now, here is where commonsense comes in. I may sound like a broken record but I'm going to harp on this issue some more as in previous chapters.

This guy knew that I was an American and knew I was coming to meet with him. If he were going to break bad, and say, blow himself up, I (the American) would be the intended target. That's not rocket science. When situations like this arose, either Sammy or one of my agents would initially meet the informant alone, in a public place. What do Afghans do when they first meet? They hug and embrace. As I've said before, it's custom to do so. Therefore, the lone agent would meet the informant and give him a huge man hug—even pick him up off the ground and shake him. The informant interpreted this as either the agent was very friendly and glad to see him, or he had been drinking Pakistani hooch. Either way, the real purpose was to determine if he was wearing a suicide vest. If he were, the agent would detect it during the hug. If the informant was packing any weapons they would fall to the ground. In case of either, the agent could choose to knock the guy out, or just save himself and take cover. The worst-case scenario was that the bad guy succeeded in detonating the vest and I lost one person. Not multiple people, just one. It wasn't that I was scared to take my turn with searching folks, but I was always the most probable target. If they saw me walk up, they had already won the game. *Why make it easy for them?*

Sammy returned a few minutes later with the guy in tow and hopped in the back with Agent M. We drove to the park at the big Mosque and found a table. A young boy about seven years old came over to take our order. We ordered some drinks, and he ran off to acquire them at the nearby shop. He didn't actually work for the shopkeeper. He was just acting as a freelance waiter in the park.

"So, what do you have for us?" Sammy asked.

The informant advised that he knew of several heroin labs and one stash house where heroin was being stored. He would give us one deal at a time to ensure he would be paid. He had tried to do work for some young American soldiers, but they had left and screwed him on his money. If we did him right, he would work for us full time. That part made me cringe, because I knew I would be the one on the hook trying to secure funding to pay him. As far as I knew, DEA (Drug Enforcement Administration) was still "moving in" at FOB Fenty and was not going to be of any assistance. I kept hearing they were coming, but months had passed. The restrictions they would have to play by would not do us any good, either. If they did get involved, the slowness of their bureaucracy would probably impede our dope deals.

Of all the deals discussed during that first meeting, the stash house seemed to be the simplest of the operations. I may have said it before, but the simpler a deal is, the more probability it has of actually going. When you get greedy, things get too complicated and you end up with nothing. I told the informant we would start with the stash house and go from there. The informant advised he would make a trip to confirm that the heroin was still at the location, and get back with us. As we parted ways, I was pretty confident the snitch would come through. He seemed cool about the deal going down, like it was child's play. He had obviously been to the labs and the stash house because the information was very detailed.

A few days later, the snitch called Sammy and told him the heroin was still in the stash house. We could take it down as soon as possible. We began our mission planning and got to the point where we needed to coordinate a raid team with the general. We put in a call to him, and found out they were about to kick off a seven-day poppy eradication operation in Khogyani District. The majority of his men were tied up with the op. The Americans were setting up a command post and had promised to provide air cover. Our gig would have to wait.

We had no other choice but to put the deal on hold, indefinitely. We hoped it would only be about a week, but that soon changed. Immediately into the ANP's operation, they met heavy resistance from the locals opposed to their poppy eradication efforts. Several men were killed, many were wounded, and several had been taken prisoner. At the time, they were only concerned with recovering their missing men and trying to retreat, while licking their wounds. While the mission had good intentions, it didn't go well. The perception among the people in Jalalabad was that the government in Khogyani no longer existed. The Taliban was the only government there. Perceptions were that the Afghan police and the U.S. military had been defeated after two days, and had to run from the fighting. In reality, the U.S. military hadn't done a thing to help the Afghans, other than establishing a command post. The Afghans probably thought they would have air cover from the Americans, but who's going to clear an aircraft to drop a bomb, during an Afghan only operation? Americans don't trust Afghans to come over for dinner, much less trust their judgment about where to drop a bomb. Nevertheless, the public perceived it as a total failure. Khogyani had temporarily been written off as its own country.

While there were many good men dead, injured, or missing, it hadn't been our fight. We knew several of the officers but there was nothing we could do at the time. Too many people were involved with cleaning up the mess. Once things cooled down, and if there were any men still missing, we would help find them.

At that point in time, our heroin raid wasn't going to happen. It would be a couple of weeks after the debacle in Khogyani, before the general had enough confidence to commit troops back into harm's way. I had suggested to him that a successful heroin raid was exactly what he needed, to recover. It would show the people that even though the ANP took a beating, they would not be deterred from enforcing the law. I still think the potential publicity from our gig would have overshadowed the negative perceptions. The general disagreed. It was somewhat shameful for him within the culture, to be unsuccessful. They just needed some time to regroup.

Against my better judgment, I thought I would go talk to the Special Forces (SF) unit that was aboard FOB Fenty, and see if they would back us up on the operation. We just needed some shooters in case things broke bad. They didn't have to do anything else, other than to roll with us as security. A simple request, but I knew better. To make a long story short and spare myself the embarrassment, the meeting ended with the sergeant major arrogantly telling me, *We can't help you.*

I decided this would be a good time for me to go on leave and recharge my batteries. I understood the Afghan general's reasoning for not backing us up, but I did not understand the American excuse. I guess the gym would have closed prior to them getting back from the mission, or maybe it was steak night at the chow hall. That's condescending, I realize, but that sergeant major was an incapable prick. All he had to do was say no, and spare the drama. Or, give us a few shooters for security, thereby helping to enable the Afghans. The SF operators would have loved to roll with us, I'm sure. They weren't afforded the opportunity because of one safety-conscious moron who was only concerned with his fitness report. He certainly wasn't going to take any risks with those untrustworthy Afghans.

The team stood down in my absence. I left them with one mission while I was away, and that was to move into the new safe house. After a lengthy negotiation and approval process, I had secured us a new crib. I'll admit, it was more like a palace than a house. But, because of all the hard work the team was doing, I felt like they deserved something nice for a change. I'm enough of a forward thinker to know that once the gringos pull the funding on the program, they won't have a safe house at all. In the meantime, we would temporarily live like kings.

●●●

A couple of weeks later, I returned. During my leave, I had received word that I was being reassigned to Kabul to take over the entire ISU program. The particular circumstances will be clarified later. My replacement, Harry, was already on deck at Task Force Paladin, waiting for me to get back. I had a short period of time to get him spun up on our operations, but I hadn't forgotten about the heroin deal.

The defeat in Khogyani had begun to fade. The move had gone well and the team was ready to get back to work. We finally got the approval for a raid team from the general, and our operational planning kicked off into high gear. We were going back into Narghosa Village, where we had pulled the hostage rescue. They would probably not be as fond of us on this trip because we were going to steal their heroin.

At 5:45 a.m., Sammy, Harry, and I met up with the Dirty Dozen at police headquarters. A couple of our agents had the informant in pocket and had already dressed him in a police uniform. The routine of coordinating the efforts of the operation was conducted by our colonel and the counter-narcotics colonel. I had little involvement. It was a good sign that my efforts and tactics were beginning to pay off. These guys were starting to operate on their own, without the gringo telling them what to do and exactly how to do it. A prosecutor was on deck to roll with us, along with the female counter-narcotics officer who was dressed in a *burqa*. With everyone loaded for bear, we rolled out.

At approximately 7:00 a.m., the raid team pulled into the Achin district center. We were joined by ten additional ANP officers in two Ford Rangers. A force of forty men headed toward Narghosa Village. As there was only one way in and one way out, there was nothing to do but hope that the bad guys hadn't placed any bombs in the road, since our hostage rescue operation.

As we rolled in, the Afghans deployed officers with PKM machine guns in overwatch positions on top of the ridgelines. By the time we got to the vehicle staging area, we had a daisy chain of machine gun positions overlooking the route.

As we began our patrol into Narghosa, the villagers gathered on the small mountain to watch us. Our patrol was uneventful, and soon we had climbed the rocky face and were in the village. The informant led the way. With his face covered, no one suspected he was anyone other than an officer. He led us straight to a residence, at which point men began to filter around to the back. A perimeter was established and held.

Our colonel made contact with an elderly gentleman who claimed to be the homeowner. The poor old guy had to be eighty years old. He had glasses that were about two inches thick and he might have had three or four teeth in his mouth. During the conversation, the old man told the colonel he didn't approve of people selling heroin. The colonel was free to search the home if he wanted to. I didn't doubt the fact the old man owned the residence, but I didn't believe he lived there. It was probably his son or some other relative who actually stayed in the home.

We began to search the small, two-room dwelling and quickly located several cotton sacks containing heroin and opium. The tally came out to be 43.5 kilos of raw opium, 1.5 kilos of processed opium, and 7.5 kilos of finished heroin. In the States, this would have been a huge heroin bust. In Afghanistan, it wasn't of epic proportion. Either way, it was the objective of the operation. Several photos were taken and the evidence was collected by counter-narcotics officers.

During the discussion with the old man, and based off of additional information from the informant, we went to a second home in Narghosa. As we began to move from the first location, we noticed that more and more villagers were beginning to gather.

When we approached the second house, several teenage children jumped out of a back window and bush-bonded up the hill overlooking the home. One of the ANP officers advised they were carrying rice bags. Apparently, once the adults realized their house was next, they dispatched the children out the back with the heroin. They had too much of a jump on us to pursue. A search of the second home netted us an additional 2.7 kilos of raw opium that had been left behind. However, we weren't done yet. As we departed the second home, I noticed that the amount of villagers in our immediate vicinity had tripled. We were drawing a crowd. I noticed that the young ANP officers were getting nervous because of the plethora of people.

Colonel G, the counter-narcotics colonel, Sammy, and I made contact with a young woman at the third residence. She didn't bother to cover her face while addressing us. After a brief conversation in Pashtu, we were in her residence conducting a search. As bags of loot were found, the woman began to cry and issue pleas to the group.

"What's she saying, Sammy?" I whispered, so she wouldn't realize I was a gringo.

"She's saying to please leave one bag for her, so that she will be able to feed her children. She said her husband will kill her and the children will starve, if we take it all."

Tears streamed from the woman's face as she held an infant in her arms. The baby couldn't have been more than three weeks old. Sammy and I looked at one another, and both knew we were thinking the same thing. She wasn't lying. There were no other opportunities for her. This wasn't the States, where drug dealers make lame excuses as to why they can't work a normal job. There were no normal jobs there in the village. Processing heroin was how they fed their children. It was nothing more than that. This young woman probably didn't even have the faintest idea of what heroin did. To her, it might as well have been flour.

As the crying turned to screams of desperation and begging in that mud house, I began to change. It hit me suddenly, like a ton of bricks falling on my chest. I was no longer driving a successful heroin raid to rid the world of an evil substance. That suddenly became a line of bullshit I couldn't even convince myself to believe. I was about to starve a family because of my efforts. As my fellow narcotics agents read this, they will probably call me a *pussy* and disagree with this assessment. Trust me, it goes against the principles that are ingrained in me from all my years of fighting the *War on Drugs,* as well. But, look at reality. We only seize about three percent of the dope in the U.S. When have you ever taken off a load to where the price of dope was affected? Never. You can take two hundred kilos of cocaine off the streets, and a ten-dollar crack rock will still be ten dollars, just one block from the federal building. This is a debate for another forum, so I'll get back on topic.

At that moment, all of my motivation and aggression for finishing the mission left my body and soul. I wanted to tell the colonel, *Just leave one bag behind, and let's go.* I couldn't take my eyes off the infant in the woman's arms. I felt dirty. Throughout my law enforcement career, I had an internal policy that I abided by. If what I was about to do would cause me to lose sleep, I didn't do it. I had done several things during my rookie year, because of policy, that doesn't sit well with me to this day. After that first year, my own conscience would override any policy that was not in the best interest of commonsense and human decency. I'm not a cruel person and could never be. But at that particular moment, I was merely a thief, stealing the woman's heroin. It equated to stealing the food from her cupboards. I could write a book on those few minutes. It opened my eyes to reality, and provided a new perspective in a way I will never forget.

The tally ended up being 2.85 kilograms of raw opium, 24 kilos of processed opium, and 32 kilos of heroin. With the woman still holding the infant and pulling at our man-jams, we got the hell out of her house. She went with us. Outside, her screaming began to rally the villagers as to the gravity of the situation at hand. They began to shadow our movements. We were severely outnumbered, and all got the sixth sense it was time to go. We had worn out our welcome.

Sammy and I volunteered to hump the rice bags full of dope, back down to the vehicle staging area. It was our deal and we felt compelled to do the grunt work, but the bags weighed on us like a ton of bricks. When I initially heaved the bag over my shoulder, I got a faint dusting of powder in my face. A few minutes later, I was feeling no pain and don't recall a lot about the hump down.

The ANP officers had grabbed a young Afghan during the investigation. This didn't sit well with about twenty villagers, who were in the middle of the patrol trying to get him back. Once we made it to the vehicle staging area, a tug of war began over this young man. To avert a mini riot, the officers let the guy go. It was the smart decision. We loaded the heroin and began to mount up. The officers who were providing overwatch of the staging area were ordered to return to the vehicles. They came down off the ridgeline as fast as they could, while carrying their PKMs and the extra cans of ammunition. It took them quite a while to traverse the rocky terrain and get down the steep hill.

Since we were one of the first vehicles into the staging area, we ended up being the last one out. No sooner than I had shut the driver's door, and put the Forerunner in drive to follow the Ford Ranger in front of me—*Badadadadadadadadadowwwww!!!!!!!!!*

Automatic fire began to crank off from the ridgeline to the left, where our overwatch positions had just been located. Apparently, the shooters moved in as soon as our troops displaced. We were sitting ducks on the low ground. As the firing continued, Harry, Agent M, Sammy, and I bailed out of the Forerunner and took up prone positions. As I hit the ground, I heard a crack. It wasn't the crack of a bullet. It was the crack of two ribs on my left side fracturing, as they crunched down on a jagged rock.

Simultaneously, the young ANP officers in the back of the truck in front of us, also bailed out and went prone. The bad guys on the ridgeline were quickly looking down the barrels of what amounted to a wall of firepower, including a PKM. As we searched for targets, the gunfire quickly ceased.

Seconds of silence followed, as we all scanned the high ground. Realizing that the trucks in front of us were on the move, the two fire teams jumped back in their vehicles. As soon I put the Forerunner back in drive—*Badadadadadadadadadowwwww!!!!!!!!!*

The assholes opened back up, as soon as we were all inside with the doors shut. I hit the gas and we started hitting bumps at full steam. I started to feel the dull pain in my side as we jostled along.

"Anyone hit?" I yelled, and was answered with three *no's*.

As we made it out of the kill zone and tried to catch up with the convoy, the four of us burst into laughter.

"That was funny as shit!" I said.

"Hell yeah, can you imagine the conversation they were having? Hey Ahmad, wait until those guys get in the vehicles and we'll light them up…OK, FIRE! Wait, Stop! Stop! They're getting out! Be quiet!

Shhhhhh! They've never done that before. They usually run. Be quiet! Holy shit, they look like they're going to come after us," Harry joked.

More laughter ensued. I'm not real sure why it was so funny, but it was. It hurt to laugh. It would take a couple of months for my ribs to heal after that little incident.

When we turned onto the paved road, everyone breathed a sigh of relief. We stopped at the Achin district center to regroup and complete some preliminary paperwork. We had accomplished our mission and everyone was going home safely. While I was proud of the team and their improving ability to pull off any deal, I felt as if we had done a dirty deed that day. I kept that feeling to myself. To this day, I cannot get the screams and the pleas of the young woman out of my head. Nor can I forget the sight of her holding an infant, while I took the only means she had to feed her family. The debate over the *Drug War* is again, not the purpose of this manuscript. However, I will never look at drug enforcement in the same light.

Our total haul that day came out to be 49.05 kilograms of raw opium, 25.5 kilograms of processed opium, and 39.5 kilograms of heroin. The heroin alone is worth four million U.S. dollars, on the streets of America. In the mountains of Afghanistan, at the beginning stages of the trade, it is worth only the benefit of being able to feed your family.

CHAPTER 14

Opposition and Suppression

The culture within the United States military, is not to allow your people the flexibility to think for themselves and solve problems on their own. I don't care what you say. The culture is that you don't do a damn thing without getting someone's permission. That someone is typically sitting in an operations center, far away from the action, and has no clue as to reality. Imagine if U.S. law enforcement acted in this manner. Nothing would ever get accomplished. A cop working the streets doesn't ask permission to stop a car or to make an arrest. He just does it, based off of training, criminal procedure, and guidelines. Dope agents don't have an operations center they report their every move to. If they establish probable cause, some unlucky bastard has to write the warrant and go see the judge. Once a warrant is signed, they formulate a plan and hit the place. Strict command and control equates to slowness. It allows the bad guys to always be one step ahead of you. While this culture may serve a purpose during a conventional conflict, with tanks fighting tanks, and infantry fighting infantry, it doesn't work when you are trying to police and influence the people. The troops on the ground need to be highly mobile, and need the authority to make their own decisions based on commonsense. That's a necessary element during a counter-insurgency operation, which is not in practice by the U.S. military. The reason is because it is not in line with conventional thought.

For an entire year, the U.S. military continually attempted to suppress our operations. While CSTC-A was funding the program, and expecting progress and results, everyone else was trying to shut us down. It boils down to the fact that Americans are only concerned with what is good for Americans. Military commanders are only concerned with what is going to make them look good, and *define* their command.

Internal bureaucracy is the focus. Enabling the Afghans to handle their own dirty work doesn't factor in. They couldn't give two cents about what happens to the Afghans or Afghanistan. Helping the Afghan government instill confidence in the Afghan people, that they are in charge, doesn't generate medals or make higher-level briefings. Firefights, Predator strikes, airstrikes, indirect fire, and body counts are what *define* military commands. In our case, having a contractor and a bunch of Afghans doing police work (without their permission and oversight) was counterproductive to them being in control of their battlespace.

It didn't stop with the U.S. military—I believe my frustration with them has already been outlined by this point. The company tried to suppress our efforts as well. During the height of my tour, the big boss from D.C. came to visit. I won't mention his real name, but we'll call him Smitty. The intent of his visit was to lock us down on the prison camps altogether, because he had heard we were rolling *outside the wire*. I explained to Smitty that the company couldn't have their cake and eat it, too. If they wanted success stories to pump out (which could potentially generate more positions, thus making them more money), I had to be with my team. I couldn't run operations from the prison camp. You can't teach, coach, mentor, and advise from the safety of a Green Beans Coffee Shop, or from the inside of a gym. If he wanted me to stay confined to the prison camp, there would be no success stories and the program would go stagnant. After arguing this point to the bone, Smitty asked me to describe a typical day, especially the part about how I traveled. How did I get from point A to point B, to run these operations? How did the military assist me, monitor me, provide support, etc.? Did I travel in their convoy? Did they have a convoy standing by for backup? I had to force myself not to laugh out loud.

"You really want to know? You may not want to know the ground truth," I said.

"Yes, give me the ground truth. Who do you coordinate with and who rolls with you?" he asked.

"No one, it's just me and my interpreter. How do I get around? I walk out the back door, get into my Forerunner, change into my manjams, and roll out the gate."

Smitty was taken aback by my response.

"Smitty, it's like this. If anyone from the military gets involved with our operations, they will screw up everything. I'll just call it a day and go home, because I'd be wasting the taxpayers' dollars."

Now, Smitty is a good man and a patriot. He has spent his entire life serving the citizens of the United States. However, he's a devout company man as well, and what I refer to as a *corporate patriot*. In the private sector, the bottom line always trumps patriotism, no matter how many flags you wave. By the end of hashing things out, I had to agree to roll in an up-armored Toyota Land Cruiser, and do a better job of coordinating with the military in case I needed help.

...The only problem was, *I didn't have an up-armored Toyota Land Cruiser. They cost about $150,000 each, and I just couldn't go and pick one up from supply. Besides, I couldn't operate undercover with an up-armored vehicle. In our area, only gangsters, politicians, and the CIA rode around in up-armored vehicles. I didn't feel like getting killed by being stupid. What the attorneys, the safety types, and the cubicle squatters in Washington don't realize, is that an up-armored vehicle only stops small arms fire. An RPG round will cut through its armor like a knife through hot butter. A small bomb in the road will mess it up as well. It's a false sense of security that makes you a huge, slow target out in the country. In Kabul, everyone drives up-armored vehicles, so it's no big deal. But out in no-man's land, you don't want to be caught riding in one. You want to be in the beat-up 1997 Forerunner I was driving at the time. Being undercover was the safest way to travel. Anonymity was my security, but conventional minds could not comprehend that concept.*

The other issue was about having to coordinate with the battlespace renter. That absolutely couldn't happen and the team still be productive. It was a waste of time as well. Even if I had coordinated our operations with them, they would not have helped us. Had we been getting our asses kicked and called for fire, whether it be close air support, artillery, or simply a couple of mortar rounds, no one was going to authorize a request for fire from a *nasty old contractor* and a bunch of *Hajjis*. An aircraft could have even been on station, but no one in an operations center was going to clear him to engage targets based off of our observations.

I liked old Smitty and I hated to deceive him, but he left me with no choice. The guidelines he laid down couldn't be adhered to. It would have shut down our operations. I believed in what we were doing. The lives we saved and affected from the shadows, trumped any useless policy that came from corporate.

Smitty didn't stick around long, because the first night he was there, we were hit with a barrage of 107mm rockets. One hit about fifty meters from my hooch and a few more screamed overhead. One even hit the brigade operations center. Smitty was on the far side of the airfield and far from the points of impact, but I think it was still too close for him. He rolled out the next day and I got right back to work.

• • •

When I initially arrived in J-Bad, the rent on our safe house had been overdue for six months. The property dealer (real estate agent) was on the colonel about it, daily. The colonel was on me about it, daily. I was complaining to Jim in Kabul, daily. There was a screw-up in the paperwork and it was supposedly being worked on. Meanwhile, we were being threatened with physical violence, arrest, and prosecution by the property dealer and the homeowner. It came to a head one day, when the owner showed up at the property dealer's office and demanded money. He then proceeded to choke the property dealer until the old man about passed out. The property dealer went around to various business owners and borrowed the money to pay the nutcase. His statement embarrassed the hell out of me, and still does to this day.

He said, "I don't understand. I have rented homes to the British, the French, the Pakistanis, the Russians, and several other nationalities. You Americans are the only ones who won't pay your bills, and you have more money than anyone. I don't understand you." He continued and asked, "How can you do business this way?"

I could only explain to the old guy that I personally paid my bills. I explained I had no control over the military folks writing the checks. I had no excuse. I did apologize on behalf of my country for the situation, and tried to convince him that most Americans do pay their bills. He wasn't convinced and will probably to this day be hesitant to rent to Americans. Every time I left his office, I felt absolutely sick at my stomach. You just don't do people that way. I don't conduct business in this manner, but I had to be the face connected to the debacle. It was terrible. The property dealer and the homeowner probably could have made money from the bad guys, by giving up our safe house. A vehicle bomb driven through the gate would have chalked up some good kills. A body count of a dozen Afghan agents and one gringo would have been a good press release for them. The situation was finally resolved due to the determination of two U.S. Navy Reserve officers, who were assigned to CSTC-A. They were able to correct the paperwork and save the day. They might not know it, but they honestly saved one of us from physical violence or death, over the damn rent on that safe house.

We had the makings of a good thing—Afghans handling Afghan dirty work. But, we didn't have money to pay informants. We didn't have money to pay our rent. We initially didn't have a contract in place to fix our vehicles. No one in the battlespace supported us.

The end of my tour with Team J-Bad came as the result of a bad decision I made while on R&R. I was hanging out with a couple of friends and several young ladies, at an undisclosed tropical location. I had a few drinks and decided to check my e-mail before we rolled out for the night. A message from Jim was waiting for me. He advised he was calling it quits and that they needed someone to take his place in Kabul. The job amounted to putting on dog and pony shows, attending meetings, and counting beans. It entailed a whole bunch of nonsense I wanted no part of. I thought I would try and act like a company man, and send a politically correct reply. It amounted to me agreeing to take the position, only if no one else wanted it. Besides, I would hate to have someone take over who didn't even know what the program was about. I sent the e-mail and would later come to realize something very important: don't send an e-mail when you're half-drunk, because you will eventually regret it.

A few days later, Jim let me know that, *Tag, you're it*. I had the job by default, because no one else wanted to create spreadsheets, build PowerPoint presentations, and deal with the bureaucracy.

When I made it back to J-Bad, the rumor had already circulated to Sammy. I remember him begging me not to go to Kabul, but I told him it was best for the overall program. I spun my replacement (Harry) up on what we had going on and broke him in during the heroin deal. I packed my trash and prepared to move. Sammy drove me to Kabul to start my new life. I almost cried when he drove away. I suddenly found myself alone, in the land of gringos at Campus Eggers.

I will spare the intimate details of this portion of the story. In short, I went to work but just didn't fit in. Living and being among the Afghans was where I felt at home. Being among the educated, higher-ranking officers that inhabited Campus Eggers was not where I belonged. I had made a huge mistake. The campus was crowded and I ended up getting stuck in a room full of young Albanian soldiers. I loved those kids to death, but they were still kids. They stayed up all night playing cards, talking, laughing, and basically doing what young men do. If I had been out in a remote combat outpost, it wouldn't have been a problem. In

Kabul, I had to endure meetings and stare at a computer screen all day. I needed to get some sleep at night. Instead of complaining, I took matters into my own hands and moved out into town. It was great. I would work all day and go home at night, just like a nine-to-five job back in the States. While I really didn't enjoy my new position, suddenly it didn't matter. It was just a job at that point and no longer an adventure. However, I was still contributing to the fight.

Fear makes money. No fear, no money. Apparently, my living out in town was a problem in the minds of some, who thought they knew better. I was promptly ratted-out to country management by people whom I erroneously thought I could trust. *You could have been kidnapped or killed, and it was a huge security breach* were the reasons cited to justify the double-cross. I will never understand why they did me in. I wasn't bothering anyone, complaining, or making waves.

But as Mullah Paul would later say, "My man, you try to see the good in everyone, but there are just some people in this world that aren't good people."

• • •

The next day, I had to go to Bagram to rehearse an upcoming brief for several VIPs from the Pentagon. After the first of several full-dress rehearsals, I stopped by the country management office. There, I ran into the big boss, who we'll call Eric. He immediately got that deer-in-the-headlights look when he saw me. He made an attempt to play it off, but was just too straight-laced to be able to fool anyone. *Great. Here we go. I'm sure this is a ride I'd rather not be on,* I thought. He told me to make sure I saw the assistant boss (who we'll call James), before I left. I had caught Eric so far off guard, that he couldn't even begin to tell me why I needed to see him. James was in the gym at the time, but he cut his workout short in order to come and talk to me. That was a bad sign in itself and soon we were face to face.

"I'm just going to come out and say it. Are you living out in town?" James inquired.

While I knew what was coming, it was still a bit shocking. I was pissed to the point I couldn't see straight.

"Those sorry, lazy, coward *motherfuckers...*"

It was all I could come up with. It sucks being betrayed by people who are supposed to be on your side. I contemplated innocence. As far as I was concerned, I wasn't doing anything wrong. I was doing my job, so why should anyone care about where I was sleeping at night? Kabul is safer than any city in America. Just compare the crime statistics. I could have forced his hand and stirred the pot for all involved. It's not what you think you know, but what you can prove. However, James was a good man, a warrior, and a true gentleman. He was well respected by everyone in the program. He didn't deserve to get caught up in my world, beliefs, ways of doing business, or thoughts on *the war*. Out of respect for James, I would not lie. Had it been Eric, it would have ended differently, because I owed no allegiance to him. I would have defended myself. I came clean with James and apologized to him for having to deal with problems I had generated. I told him I had a high-level briefing coming up in a few days, and that I had to perform. Whatever was going to be handed down the pipe needed to wait until after the gentlemen from the Pentagon had left. James agreed. I would focus on the briefing as if there was nothing wrong. Immediately following the briefing, I would report to his office to learn my fate, like a gentleman.

The drive back down to Kabul that night was surreal. I was about to get the boot and I knew it. I had to continue to prepare for the briefing as if nothing was the matter, knowing all along that five minutes after the mission was accomplished, my life would be altered by corporate. Over the next few days, that's exactly what I did. On the day of the briefing, I drove to Bagram, went through the ridiculous two-hour hassle of getting my vehicle searched at the gate (all the while having an AK-47 and a backpack full of frags on my person), put my personal predicament aside, and got ready to put on a good show.

● ● ●

The briefing was in reference to all the Joint Task Force Paladin programs, which were being implemented to combat the IED problem in Afghanistan. The guest of honor was Dr. Ashton Carter, the Under Secretary of Defense for Acquisition, Technology, and Logistics. At the time, I really only knew that he was an important man from the Pentagon. Accompanying Dr. Carter was Lieutenant General John M. Paxton, Jr., United States Marine Corps, who was also Director of Operations, J-3, The Joint Staff. There was a one-star army general and a slew of U.S. Army colonels within the entourage.

It was hot as hell that particular day, and the fatigue from travel showed on their faces as they arrived at the compound. After a quick briefing was conducted outside by an Australian EOD tech, the entourage shuffled into the air-conditioned class room. Everyone took their seats and settled in.

What immediately struck me, was that I felt honored and humbled to even be in the same room as men of their caliber, education, and integrity. To be honest, I didn't feel like I was supposed to be there. I was just a lowly, uneducated, narcotics agent from Atlanta, Georgia, who had been running with a bunch of Afghans out in Nangarhar Province. And although we had saved numerous lives, we had ignored every policy, law, and restriction imposed by the United States government. Yes, I'm sure that I should not have been in that room.

When it was my turn, I briefed several of our cases and showed pictures of the unlucky bastards we had rolled up, along with pictures of their contraband. I kept it basic so everyone was clear on how we did it. It was accomplished by simple, undercover, proactive, police work. I made no attempt to hide the way we operated. I figured I would let the results speak for themselves, and let the chips fall where they may. The only thing I was political in defending was the issue of running informants. I reiterated the company line that we merely *advised* the Afghans on how to run *their* informants. While I was nervous as hell initially, once I started talking, I relaxed. Besides, I was probably getting canned in about an hour anyway, so it really didn't matter. Dr. Carter thanked me and my men for our efforts, as I concluded my brief and sat down.

As the brief progressed, I firmed up my initial opinion of Dr. Carter. This was one of the smartest individuals I had ever met, but he was also a down-to-earth guy. I knew nothing of his background, but knew he was not just some suit who had been appointed to his position for political reasons. It was obvious he had the credibility to back up his words. I was surprised, because most of the upper-echelon government types I've met are phony. They don't know what the hell they're doing because they don't have the background or credentials to hold their position. They portray that they do, but they don't.

After the brief was over, everyone snapped to attention as the honorable men stood to leave. Dr. Carter shook hands with several folks in the room and started to walk out. LTG Paxton made a special beeline for where I was standing, and shook my hand.

"I just wanted to tell you I appreciate what you and your men are doing, and I appreciate you volunteering to come over here."

"Thank you, sir. I'm glad to be here," was my response in a broken voice, as I almost shed a tear in front of him.

The general then filtered out of the room behind Dr. Carter. I felt a bit out of place, because the general made a point to talk with me specifically, without acknowledging any of the other speakers. I wondered why he made it a point to shake my hand. Maybe he has a son who is in law enforcement. Maybe he appreciated the simplicity of what I had to offer and how I communicated it. After all, they were used to being briefed by people talking academics, with stars on their uniforms and PhD after their name. Or, maybe he just felt sorry for me because he knew I was running with the Afghans. I don't know.

That was the first time I felt like all of the risks we had taken, all of the opposition we had endured, and all of the bullshit we had put up with, was for a purpose. In our hearts, we knew we were saving lives but no one else seemed to acknowledge that. Our actions were always met with threats, opposition, and degradation. But in that room, an undersecretary of defense and a United States Marine Corps lieutenant general had acknowledged our efforts, and said thanks. Suddenly, all of the issues I was about to face really didn't matter anymore. I had just been face to face with men whom I considered to be true patriots. They weren't trying to impress anyone and climb any corporate ladders, because they were already at the top. I think it's why they seemed so focused on the mission and why they appreciated ground truth. While I had nothing in common with them outwardly, I felt that inside, we were all on the same sheet of music and were thinking the same thing. *Enough of the skewed statistics and dog and pony shows. Just prevent some of these bombs and quit arguing over numbers and words on a slide.*

● ● ●

True to my word, I was down at James' office, five minutes after the meeting. I was on top of the world, and wasn't the least bit concerned about what the *corporate patriots* at Company X had come up with. James told me to go back to Kabul, and that he would call me with the verdict from the powers that be. The next day, word came down that they were removing me from my position within the program. I was being transferred to an obscure position with an army unit somewhere.

"Respectfully, may I have tonight to think about it?"

It wasn't what Eric was expecting. He thought I should be thankful I wasn't being fired, accept the transfer, and shut the hell up. He may have been concerned with the bottom line as well, because the company loses money when a person leaves Afghanistan. It behooves them financially, to try and keep people in place, regardless of the circumstances. *Just conform and go with the flow. We'll all make money.*

"I'll let you know what I decide by tomorrow. This may be a good time for me to go home for a while."

I thought about it that night, and decided it was time to go home and decompress. The next day, I told Eric I was respectfully declining the transfer, and was popping smoke and rolling out. I would pack up and get to Bagram within the next few days.

When I finally got to Bagram, Eric and I spoke. He shook my hand and expressed his respect for me that I was taking my medicine and departing peacefully. It seemed sincere, but I have to concede that maybe he was just trying to keep me quiet until I was gone.

"Sir, I'm the captain of my own destiny and I take responsibility for my actions. That's just who I am and how I was raised. No need to stir up anything. Besides, it's time for me to take a break. But, just do one thing for me, if you don't mind?"

"Sure, if I can," he said cautiously, as he had no motivation to do anything for me at that point.

"Just let me leave with the reputation of being the guy who was removed for living among the Afghans. In six months, it will only add to my credibility. In a couple of years, the reputation will be worth even more. No need to hide it, because I'm not ashamed of it. The Afghans are good people and I have made many friends here."

Eric had never looked at it in that light. He was an intellectual, so he appeared to be evaluating the statement objectively. I think he knew I was right. It was, and is, a fact that I was removed from my position for living among the Afghans—the same folks I was being paid a quarter of a million dollars a year to advise. The irony is uncanny. But, *no fear, no money.* I surmise that my living out in town could have portrayed that Afghanistan was no longer dangerous. Therefore, people might lose their combat/hazardous duty pay.

● ● ●

I left Afghanistan with no hard feelings, no shame, and no bridges burned. I would miss Sammy, my team, Jalalabad, Kabul, and Afghanistan as a whole. It felt like home to me. I made it to Ali Al Salem, Kuwait, where it was hot, as usual. I was traveling with several other guys leaving the program, and soon we were settled into a nice air-conditioned tent. I grabbed my laptop and figured it was time to break the good news to my wife, that I was finally coming home. Upon breaking the good news to her, she broke the good news to me that she wanted a divorce. She would be moved out of the house by the time I got home. I was a bit shocked at first, but I'm a realist. I had been away from home for over two years. Her life had moved forward without me, while time stood still in my mind. It is a typical story for military personnel deployed or contractors working in this business.

I mourned the situation for a good ten minutes. OK, maybe it was only five. Then, a feeling of total liberation came over me. I was about to be free from being deployed and I was going to be free from an ungrateful, white woman. If there had been a bar on Ali Al Salem, I would have bought the whole camp a drink. I was ecstatic with excitement, in a twisted sort of way. I couldn't wait to get back home and start straightening things out.

While I may never know the real reason she wanted a divorce, I do have my theories. In my absence, my wife had grown closer to her faith. Already a devout Christian, she had consumed herself with the church. Her dedication had turned cult-like. It was almost as if it were an addiction. I was no longer speaking to her as my wife, the individual, I was speaking to a representative of the church.

My involvement with the Muslims in Afghanistan had cast doubt in her eyes, as to my true religious beliefs. *Once again, religion divides.* She wasn't alone. In addition to my wife, several family members and friends perceived I had converted to Islam. It was comical, but understandable from their perspective. They were all unaware that the bracelets I wore *religiously* (no pun intended) on my wrist, were associated with Buddhism. While in Thailand on R&R, several little-old-ladies and a Buddhist monk tied those bracelets on my wrist, and blessed me to keep me safe.

Looking back, I'm not really sure who or what kept me safe during my travels and adventures in Afghanistan. I had several churches and hundreds of Christians praying for me in the United States. From Thailand, I had Buddha, a temple full of monks, and a plethora of eighty-year-old ladies looking out for me. In Afghanistan, I had an entire homeless camp praying to Allah to keep me safe. I would imagine those folks still pray for me today. So, what entity can claim credit for my safe return? Maybe it was just the ritual that Sammy and I did before each operation. I will never know.

CHAPTER 15

Afghanistan Tourist Visa

After spending a month roaming around Thailand and Cambodia while working on this book, I felt the need to return to Afghanistan. I wanted to clarify several points with Sammy to ensure I wasn't missing anything. I also wanted to be able to put boots on the ground, without being under the rules and regulations that accompany working for the U.S. government. It would be a pleasant change, to not have to answer to any gringos and just visit with my Afghan friends. I booked a ticket to Dubai, and soon found myself sunning by the rooftop pool at the hotel. The next morning, I got up and hopped a cab to the Afghanistan Embassy. Upon arrival, it immediately felt familiar.

I walked in and realized I was the only gringo in the building. I handed my passport and visa application to the lone gentleman at the first window. He made a copy of my passport and stapled my photos to the application. For this service, he charged me fifteen U.S. dollars.

He directed me to the second window, where five men sat in front of their computers. After filling out a receipt, the young Afghan told me to go to the Azizi Bank window, and pay fifty U.S. dollars. I accomplished this task and returned with my stamped copy.

"How many days can I get?" I inquired.

"Thirty days," was the response.

"I can't get the multiple-entry for 180 days?"

"You cannot. Come back after three."

That was the extent of my conversation with the visa kid. It wasn't that I had planned on staying that long, but it made sense to try and get the 180-day visa, just in case. You never know where the wind might blow you. That philosophy is especially true, because of my principles pertaining to adventures.

At three in the afternoon, my passport was ready. I now possessed a tourist visa for what the rest of the world considered a war zone. I went from a heavily armed operator to a mere tourist. I laughed, as I looked at the sticker on one of the last pages in my passport.

My flight was scheduled to depart at 3:30 a.m. I didn't bother trying to sleep the night before, and arrived at terminal one by 2:00 a.m. I had plenty of time to spare. I wasn't familiar with terminal one, and asked one of the airport employees where the check-in counters were located. He directed me through a set of double doors to departures. All of the check-in counters were closed, which was to be expected at two in the morning. I would be first in line.

Thirty minutes later, no one had arrived to man the counters. As a lone airport employee wandered through, I again inquired as to where the Safi Airways check-in was located.

"Next building over," was the reply.

It was customer service at its best. I had come early to purposely avoid having to rush. I scurried over to the next building, and caught the only Safi Airways employee in sight, as he closed down his terminal.

"Is this the Kabul flight?" I asked, while out of breath.

"Where have you been? We have to close one hour before. Why you wait so long?" he asked.

Shit. I looked up at the screen and noted the time was 2:33 a.m. It was three whole minutes past his deadline. It seemed as if I would not make my flight as planned. I stood there and stared at the heavy-set Arab while still half-drunk from being in the hotel bar all night. I was getting frustrated at his laziness. He reluctantly logged back into his terminal and printed my boarding pass.

"You have to hurry. It is a long walk to the gate. It's the last gate and then you have to board the bus."

With that, I began about a mile-long journey, rife with obstacles at passport control and security screening. I was running through the airport, O.J. Simpson style. Sweating profusely, I made it to my gate as the last person in line was checking in. We immediately boarded the buses on the tarmac.

The American pilot made a few announcements and indicated the flight time was two hours and ten minutes. It took about two weeks to get to Afghanistan via the U.S. military procedure. I could deal with a couple of hours. I looked around the plane and took note of the makeup of the passengers. There were a few federal agents of some sort. You can easily pick them out of a crowd. If they're not wearing the tell-tale 5.11 pants, just look at their shoes. They'll probably be wearing a brown pair of Merrell's. They will constantly be scanning for threats. I continued to look around and spotted a couple of Afghan kids traveling with grandma. *Kids? Traveling to a war zone with grandma?* Everyone else on the plane was a gringo, who obviously worked for the U.S. government in some form or fashion. Some were veterans to Afghanistan, while others you could tell were a bit nervous about making their initial trip.

We touched down in Kabul, right on time. Due to the layout of the airport, the plane had to be backed into a parking spot. As we boarded the buses to the terminal, a group of Afghans began wrestling our bags off the plane. While there appeared to be one gringo in the mix, directing operations, there was no sign of any U.S. military personnel on the commercial side. It was comforting to me, that at least we were letting the Afghans try to run something without our intervention.

Passport control consisted of several booths, but only three were manned. After a quick stamp on my tourist visa, I was waiting for my bag. Thirty minutes later, the belt started turning. I grabbed my bag and walked the twenty feet over to the x-ray machine that served as customs. A couple of minutes later, I was free to roam about Afghanistan.

In the words of Doctor Martin Luther King, Jr., *Free at last! Free at last! Thank God almighty, we are free at last!* I was no longer tied down with General Order Number 1 and military regulations. I was no longer a contractor for the U.S. government. I was merely Mark, the *tourist*.

A quick call to Sammy confirmed that he and a friend, who we'll call *Engineer*, were waiting out in the front parking lot. The ANP wouldn't allow them to come to the terminal and wait. I knew that Sammy had to be pissed because some lowly officers were telling him he couldn't enter the secure area.

"As-Salamu Alaykum, my friend!"

"Hey bro, it is good to see you!" was Sammy's reply.

After hugs and handshakes, we were off in Sammy's vehicle. We traveled to the home of my friend Farid, who was also Sammy's first cousin. Farid was a candidate for parliament in the recent elections. While the voting was over, the ballot counting was still in progress. I had been following various news reports over allegations of fraud and mismanagement of the election. It was the topic of the week for international news outlets. While everyone was fairly confident Farid had been elected, it wasn't a certainty at that point. Fingers were crossed.

We arrived at his home and began to catch up on things. Farid had been campaigning non-stop and was exhausted. He was glad the elections were over, whether he had been successful or not. Regardless, his two cell phones were constantly ringing. We weren't able to talk about much because he was inundated with phone calls.

"Let's go. Some of my friends have made hospitality for us on the way to Panjshir," said Farid.

With that, Farid grabbed the keys to his Land Cruiser. Sammy and Engineer sat in the back, while I rode shotgun. As an American, I am used to always knowing what the plan is. When I go anywhere in the States, I know where I'm going, who I'm going with, what I'm going to do when I get there, and how long I'm going to stay. It's just the way we do business in our culture. I had long before learned, this was not the Afghan way. Just get in the car and go. Don't ask questions. You might get back in an hour, or you might come back in two days. You may be meeting two people for lunch, or there may be a hundred. That's the gamble. The bottom line was, I never knew what kind of forum I'd find myself a part of during my travels in Afghanistan. That aspect actually made for a more exciting adventure.

We headed out of town on the road to Bagram. Farid pulled over to the side of the road and stopped, after receiving a call. A Lexus SUV passed by and honked the horn. It had five people in it. Farid explained they were his friends, and we would follow them to the location. We soon turned right onto a dirt road, which led through several sparsely populated villages. An hour later, we arrived at a rather large qalat with a nice steel gate. As we entered, we were greeted by at least twenty people.

I'm rather skilled at fooling even the most rural Afghan with my appearance, limited language ability, and my mannerisms. I was doing OK with no scrutiny, until Farid told the group I was his American friend.

"America? This guy is from America? He looks Panjshiri."

"No, he looks Nuristani."

"No American has ever come way out here."

I became an instant celebrity. My low-key personality had to shift from introvert, to Mr. Politician. I needed another beer.

...I enjoy a good party, don't get me wrong. The thing that is difficult for me, is when I get the royal treatment. I'm not used to that. When Afghans would find out that I was American, I would get the best seat, the best food, and have every need attended to. A poor man like me isn't comfortable being catered to. I can't stand four and five-star hotels. While most members of the

U.S. military think the Afghans are all out to get them, that is the farthest thing from the truth. On a personal level, they love Americans. They always treated me like I was a rock star.

The host was in his late fifties. As the group wandered toward the house, he asked me if I wanted to sit inside, or in the garden. As I didn't want to make the decision for the whole group, I told him I was fine with either. Once he realized I wouldn't give a firm answer, the host elected for us to stay in the guest room upstairs.

It was very nice with a beautiful view of the acres and acres of grape vines. Since alcohol is technically against the principles of Islam, there are no wineries in Afghanistan. All of the grapes are grown and sold at the market. It's a shame, because the place would have made a very popular and beautiful winery. I had spent a lot of time in California's wine country, but none of the vineyards in California compared to where I was standing.

● ● ●

As they went around the room introducing people, I realized I was hanging with political clout. The three-letter intelligence agencies undoubtedly were listening to some of the phones in the room. There was another gentleman who had just run for parliament, and several high-ranking officers with the Afghan army. One older gentleman, who came across as a comedian, was reportedly, the best friend of President Karzai's brother. I'm not exactly sure who another older gentleman was, but he offered to take me on a tour of the Taliban camps in Helmand Province. Upon his personal guarantee, no harm would come to me if I wanted to go. A lot of people talk shit, but he was serious. In response to this offer, the only objection the room had, was that in general, you can't trust the Taliban. Besides, no one else in the room could imagine why anyone would want to go visit the Taliban in Helmand, anyway.

We spent a wonderful afternoon with the gentlemen. Most of the conversation was jovial. Storytelling, jokes, and politics were discussed between the laughter. A meal fit for a king was served in an adjacent room. I ate until I couldn't handle another bite. The day turned out to be quite an adventure, and I was glad I had been invited.

Afterward, we indicated to Farid that we needed to roll. It had become obvious he would have stayed all night, but we had a three-hour drive to Jalalabad. As we said our good-byes, one of the older gentlemen took off his turban and put it on my head. Everyone laughed, and about twenty people armed with cell phone cameras began taking pictures of me. That particular turban was rather extravagant, I must admit. A conversation about turbans immediately ensued as I stood there modeling my new attire. The gist of it was, the type of turban worn is merely a style. It doesn't indicate that you're a member of the Taliban if you wear a certain turban. This contradicts the stereotypical image of what some Americans perceive to be Taliban. If you wear a black turban in this fashion, you must be Taliban. That's not accurate. It's kind of like saying everyone in America who wears a red Chicago Cubs hat, is a member of the Crips. It is merely stereotypical and not factual.

As we passed by Masood Circle in Kabul, Farid told the driver to pull over. There were several men eating at a restaurant whom he needed to talk to. He was soon sitting at their table while we waited in the car.

"I don't think Farid understands we need to go," Sammy uttered.

"He's in politics now, my man. It's all about shaking hands and holding babies. That's his job. But you're right, it's getting late and we need to roll."

We got out of the vehicle to stretch our legs and were approached by a heavy-set woman who was a street beggar. Sammy tried to shoo her away, but she just started making jokes about him. She was quite the character and ultimately, Sammy pulled out a five-Afghani bill, and tried to hand it to her.

"Who can benefit from that little bit of money?" she joked, as she pulled out five, ten-Afghani bills and showed them to Sammy.

Sammy laughed and told her she was crazy.

"Take it or don't. It's up to you," Sammy chuckled.

After a few more jokes, the woman took the five-Afghani bill, and began to walk away while spewing comedy. There's no telling how much she made on a daily basis, because she was very entertaining.

Finally, Farid realized we were waiting on him. He ordered his driver to take us to the house so that we could leave. We said our goodbyes and wished him good luck on the pending results of the election.

The drive from Kabul to Jalalabad is beautiful. It can be congested at times, trying to transit the tunnels that lead through the mountain pass. One wrong turn in this area, and you and your vehicle will careen off a cliff to your death. In the past, we usually made this trip with a trash bag full of ice and Tuborg beer. It's easier to face the treacherous route when you've been drinking.

"Hell, go ahead and hang a hard right," we would joke, as we drove through the pass. That hard right would have meant death.

It was a different feeling once we arrived in Jalalabad. I didn't have a prison camp to go home to. No chow hall access. No Green Beans Coffee. It was just the local economy now. No huge paycheck was accumulating. It was just me and my friend, Sammy. My thoughts wandered back to the army major who told me I trusted *these people* too much. *Screw him, his prejudices, and his elitist attitude,* I thought. I don't think I'll ever forgive that particular gentleman for disrespecting my friends and perceiving them to be less than his equal.

It was arranged that I would stay at Farid's home in Jalalabad. He had a couple of his men living there who were watching over the place. Engineer would stay and crash out as well. Upon arrival, they were very kind and I was ushered to one of the rooms, where mats and pillows lined the walls. It was a nice guest room. As I went to inspect the bathroom, I had to laugh. It was typical Afghan style. Number one, it was dirty as could be. It had a bathtub, but no one had set foot in it in years. It had a western style toilet, but there was no water hooked up to it. Footprints lined the seat where your ass was supposed to go. You had to manually flush it, by scooping water out of a nearby barrel. There was about a half an inch of water on the floor, with a nasty pair of community sandals to use while you did your business. The sandals were the kind that squished when you stepped into them, from the years of soaking in a water and urine combination. At least the cold water on the sink worked. While it was typical of Afghanistan, it was far from the accommodations I'd gotten used to over the past month. I had gotten a bit soft while living the good life in Thailand. Really, the place was wonderful and I was thankful to Farid for the hospitality.

As I lay down, exhausted from a lack of sleep, I discovered something pertinent was missing. That something was a woman. I quickly realized, I no longer wanted to voluntarily live without the company of the female persuasion. General Order Number 1 was once again, upon me. The bad thing was, this time I had imposed it on myself.

While I had planned to stay for a month, I lay there re-evaluating my situation. I wouldn't be staying for a month, after all. It would be a short visit, as if I were visiting family. My friend's dad always said that family members coming to visit are like fish. They're OK to lie around your house for a few days, but after that, they start to stink. This would be that type of trip.

I spent the next few days with Engineer at his office, while Sammy went to work. This worked out well and set the proper atmosphere daily, in which for me to write. The office was downtown and was pretty nice. It was like a small suite, complete with a little kitchen and a bathroom that had a shower. But, there were more people coming and going than the driver's license bureau. How in the hell could they accomplish any work with all the traffic, gossip, and visitors?

In Afghanistan, when someone walks into the room, everyone stands up to greet them. The newly arrived has to hug and shake hands with everyone in the room. By the time ten people have gathered, it literally takes three to five minutes to properly conduct a greeting. To make matters worse, you have to repeat this process whenever someone elects to depart the room. It's the culture and it shows respect, but it impedes the heck out of progress. Paperwork cannot be generated, thoughts cannot be formulated, and plans cannot be drawn up, with an open office of this type. Was it networking? Absolutely, it was. Networking and relationships are the name of the game. However, there was no time to generate paperwork for contracts, bids, or proposals, like the Americans expected. Engineer had a computer and a printer, but I never once saw him turn it on the entire time I was there.

In a typical Afghan office, the television is always on. Cell phones are always on the loudest setting, with some type of American music as the ring tone. That's normal to them, but it's difficult for an American like me to concentrate with all that noise.

As I typed, I listened to the day's conversation, picking up on the few key words I understood in Pashtu. Apparently, there was some concern over a road construction project that an Indian contractor had been awarded. The gentleman was afraid of a member of parliament for some reason. I'm not sure if it was over bribes, but it obviously had to be. The Indian gentleman was afraid, to the point he just wanted to sub-contract out (basically sell) the entire contract. A discussion ensued over the gentleman in parliament. I recognized the guy's name from the past year's operations. He was a gangster and a smuggler who had gotten himself elected. Most of the members of the Afghan parliament were former warlords, in some form or fashion.

Sammy assured the crew that it wouldn't be a problem. He would go talk with the man himself, to make sure there were no issues. Sammy told them to assure the Indian man they were interested in doing the work, because there was money to be made.

● ● ●

I wasted no time in changing my flight back to Dubai. I explained to Sammy I would be calling him frequently over the next couple of months, to clarify and discuss past events. Using Skype is fairly economical, and cell phones in Afghanistan don't charge minutes for incoming calls. It would be a better arrangement than having me lie around for a month, stinking up the place. I tried to explain the fish theory from my friend's father. A week made me a welcome guest. A month made me a free-loading brother-in-law. Either way, I would be back. That would always be certain. We would remain friends until the end of our days.

● ● ●

They say that you can never go home again. That's how I felt as we left Jalalabad and headed back toward Kabul. It just wasn't the same. After we passed through the tunnel at Darunta, Sammy received a call from the new advisors of the team. They wanted to know if the team had any information on a female aid worker (Linda Norgrove), who had apparently just been kidnapped in Kunar Province. Sammy made a few quick calls, and discovered that our agents were already on the case. They had several snitches up in Kunar, who knew exactly where Ms. Norgrove was being held.

Sammy called the advisors on FOB Fenty to give them the information, which included the name of the person's home where Ms. Norgrove was being held. The obvious question was asked.

"No, they do not have a grid. We've been doing this long enough to know you want the grid. It's not an area where they can just walk up to and get the grid. I've given you her location, so now get in the airplane and go do some action."

With that, Sammy ended the call and looked over at me.

"Do you want to stay? I can turn around and we can go release this lady. We know enough people up there. We can do this."

He knew I wanted to go. He knew we could pull it off. It just wasn't my rodeo any more.

"The gringos are involved, so you know it's already screwed up. If they weren't involved, I'd stay. But now, it's just too complicated. You know that," I said.

That was the decision I convinced myself I had to make, because I was in Afghanistan on a tourist visa. I no longer had the authority to carry weapons or to advise any law enforcement efforts. But to be honest, those were just excuses. Let's face it, I've never been one to be concerned with policy. The real reason was what I had cited to Sammy. I didn't want to have to interact with the U.S. military or explain to them about why I was involved. I just couldn't take another beat down from my own people, especially when I wasn't getting paid and wasn't supposed to be there. If anyone found out I was even in Afghanistan, it would jeopardize any future employment with Company X. I was already suspect, because it was public knowledge that I had developed personal relationships with the Afghans. I was on the borderline of not being trusted by the suits in Washington. They could easily pull my security clearance, which would mean my livelihood. Self-preservation told me to stay the hell out of this one because too many gringos were involved.

This was a decision that on October 8th, I would come to regret. I had underestimated just how involved the gringos would become with this incident. The powers that be, within the U.S. government and military, ordered a hostage rescue operation, to attempt to free Ms. Norgrove. Reportedly, the U.S. Navy's SEAL Team Six, supported by U.S. Army Rangers, conducted a pre-dawn raid on the compound where Ms. Norgrove was being held. During the raid, she was killed. While it was first reported that an insurgent detonated a suicide vest, a subsequent investigation revealed that a Navy SEAL had thrown a grenade, which unintentionally and inadvertently, killed Ms. Norgrove. I wasn't there and was not privy to the investigation. This information is what was reported by various media outlets.

I don't think anyone should blame the young operators for the outcome of this incident. They were merely trying to survive the situation they were injected into. I do blame Ambassador Karl Eikenberry, General David Petraeus, and the other conspirators who were trying to be heroes in the press, get themselves promoted, or who were chasing medals. While this deadly raid was being planned and pushed by the Americans, there was a contingent of Afghan tribal elders who had assembled in Kunar Province. They were in the process of negotiating with Ms. Norgrove's captors, for her release. Had I stayed, we probably would have been involved with this contingent. I would surmise that the U.S. military (as usual) put no faith in the Afghan effort and attempted no coordination. *Never trust an Afghan* is ingrained in their mentality.

Besides, what would be more spectacular, more *Hollywood*, more heroic? A bunch of nasty old Afghans negotiating her release, peacefully? Or, a hostage rescue operation, complete with troops fast-roping out of helicopters, a dramatic firefight and explosions, close air support on station, and Predator aircraft loitering overhead airing the footage live? From the American perspective, it's obviously the latter. That mentality cost this young woman her life, unnecessarily.

I personally feel that no discipline should have come to the young operators who were there, in the line of fire. What happened to the theory that the captain of a ship is responsible for everything that happens on the ship? Apparently, it doesn't apply in this case. It sounds eerily similar to the outcome of the My Lai investigation and subsequent prosecution. Just kill the careers of the lowest ranking individuals and let the elitists go on to their next screw up.

The young SEAL who threw the grenade didn't ask to be put in that situation. He was ordered to go. He had no choice. Therefore, out of logic, I have no choice but to believe the ultimate responsibility rests with Ambassador Eikenberry and General Petraeus. They should have been disciplined for using poor judgment. They were the captains of the ship, and could have elected to be smart about the situation. In essence, they played a game of checkers when they should have been playing a game of chess. I do wonder, if Ms. Norgrove had been their daughter, would they have chosen the same course of action? If she had been my daughter, I would have preferred to exhaust *all* attempts at a peaceful negotiation, before anyone went in with guns-a-blazing.

The Afghan tribal elders would later say they were en route to the compound, when the fighting started. I obviously cannot say for certain that the elders could have successfully negotiated Mr. Norgrove's release. I cannot say that my Dirty Dozen could have even been a factor, had we become involved. But, we had become pretty good at negotiating with the Taliban and the bad guys, when it was the smart thing to do. In this particular case, talking and negotiating was the smart thing to do—not military action. That should have been plan Z.

I will forever live with the fact that I didn't tell Sammy to turn the car around. To Ms. Norgrove's family—I apologize for compromising my morals and principles in order to save my security clearance. I should have been knee-deep in the situation from the start with my Afghan friends, even if it meant getting prosecuted in my own country for interfering with *their* operations.

...Why would I pour salt into this wound? One of my colleagues asked, "Why in the world would you talk about this incident? Why would you call out General Petraeus and Ambassador Eikenberry, and directly blame them for that girl's death?" Another friend, who happens to know General Petraeus personally, argued, "General Petraeus is a good man." After the feedback, I felt compelled to explain my philosophy in greater detail, as there are several learning points that can be derived from this incident.

Since the U.S. invasion of Afghanistan, there have been thousands of innocent civilians killed by the U.S. military. They are often referred to as *collateral damage* as if they are objects, and not real people. During my time in Afghanistan, this obviously continued to occur. At first, it wasn't personal, but still sounded barbaric when someone would say, *Oh well, that's war,* after an innocent civilian was killed. When these types of incidents affected people I personally knew, it hit home. All of a sudden, *Oh well, that's war,* sounded almost criminal, as if they were getting away with murder. Even if there was an investigation and the Americans were found to be blatantly at fault, rarely was anyone held accountable. The best a poor Afghan could hope for, was a small payment as a show of condolence. Money does not bring back a loved one.

I will make the observation and the argument that the talk of a *small amount* of collateral damage (expressed as a positive thing by military planners), still translates to the *murder* of innocent civilians.

I began to resent the lack of concern for Afghan life on the part of my fellow Americans. To qualify this, I ask you what would happen if an errant artillery round hit a mini-van, and killed a family of five, on a road near Fort Bragg, North Carolina? Would we just sit back and say, *Oh well, that's just training for war?* I don't think so. There would be an investigation to no end. If anyone could possibly be held accountable, you can bet your ass they'd be prosecuted. The lawsuits would never stop. An entire chain of command would be relieved. But, for some reason, if we leave the borders of the United States, a totally different mentality is applied. A minivan full of Afghans doesn't carry the same importance as a minivan full of Americans. Doesn't that border on the edge of racism?

I brought up the incident in Kunar, because discussing Afghans getting killed doesn't spark anyone's attention in the West. If someone makes a decision or orders an operation that kills some innocent Afghans, the reaction is, *Oh well, that's war. Who cares about a bunch of Hajjis, anyway? It had to be done. They were just collateral damage. We'll do better next time.* The perception is that Afghan civilian deaths are just the price of doing business. A Captain even told me one time, *They're savages and about the dumbest people on the face of the earth.*

However, when you talk about a white, British girl's death, and blame it on an American general, suddenly it becomes serious. I don't think General Petraeus sat back and said, *Oh well, that's war.* The incident was studied, quarterbacked, and picked apart to see what went wrong. Apologies and condolences were made at the highest levels of government. Why, on this particular case? Because it was a white girl. A British girl. An English speaking girl. I have not heard one person refer to Ms. Norgrove as collateral damage. Nor should they, because it would be an awful reference, with an absolute lack of respect for her family.

Nor should we ever refer to innocent Afghan women and children as "collateral damage." Nor should we utter the phrase, "Oh well, that's war," when they are killed. That is the moral and point of this story. Afghan civilians should be valued as much as the Americans, the British, or anyone else. People who don't speak English should be valued as much as those who do. That is the difference between my perspective, and that of the leadership of the United States military.

The My Lai prosecution set a dangerous precedent. Everyone from Captain Medina on up, basically got away with murder. Only the young Second Lieutenant William Laws Calley took the rap. While he avoided spending life in prison, it ended his career and caused him to live a shadowy life. But, much like the young SEAL who threw the grenade, Lieutenant Calley wasn't in My Lai 4, on March 16th, 1968, on his own accord. He was ordered to go. At what point are we going to start holding the policy makers accountable, and not the operators who have no choice but to follow flawed orders? That is the other moral to the story.

The reality is, the situation was a kidnapping of a British citizen, and not a member of any military. It was a crime. A crime committed in a sovereign nation. A crime that should have fallen to the Afghan police to handle, and not to a foreign military force. Let's have the attorneys weigh in and quarterback this one. I'm sure they'll have something to say about the actual legality of the operation. Let's be sure to choose attorneys outside of the beltway, who can remain objective.

● ● ●

The traffic was terrible in Kabul, as usual. I ended up making it to the airport with only thirty minutes to spare before my flight. The ANP officers stopped us, and told Sammy and Engineer they had to walk to the parking lot. For some reason, only the driver could be in the car. The officer then opened the driver's door. I quickly deduced that he meant for me to get out. When I stepped out, he patted me down looking for weapons. Satisfied, they let me pass while Sammy and Engineer began humping it toward the lot. I arrived at the second checkpoint and had a short conversation with the officers there. My end of the conversation consisted of just saying, *saysh*, which meant yes, and *puishay*, which meant I understood. After parking the car, it was a sprint to get to the security line.

As I was still wearing my man-jams, the security personnel couldn't figure out why I didn't understand what they were telling me to do. I placed my bags on a table and was patted down a second time. After a brief hand search of one of my bags, it was time to put them on the x-ray belt. As I passed through the metal detector, I realized there were three people observing the screen attentively. They didn't like my roll-aboard, and I was directed to another hand search. *These guys need to come over to the States and teach TSA personnel how to do the job,* I thought. The Afghan crew was thorough and serious at what they were doing. There was absolutely no way to get a weapon past them.

After getting both of my bags searched by a customs official, I was clear to proceed to check-in. I approached the Safi Airways counter where there was no line. I knew right then I wasn't getting on the flight. I was late. The plane was still there, but they had closed the check-in process. However, this was Afghanistan. I would try to bribe my way through. After all, the world media and the U.S. military report that everyone in the country is *corrupt*. I approached the young Afghan sitting behind the computer terminal. He had on a dress shirt and tie, and had neatly groomed hair. He looked very professional. Not a good sign when you're trying to offer a bribe. I told him I was sorry for being late and needed to get on the flight back to Dubai. I slid a piece of notebook paper onto the desk, which had my confirmation number written on it. Cleverly, I had placed one thousand Afghani under the piece of paper. The corner of the bill was clearly visible.

"We've already been instructed to close, but I'll check with the supervisor just in case," said the young man.

With that, he placed a cell phone call, but was told by the supervisor that it was too late. The flight was closed. He hung up the phone and apologized to me for the situation.

"Are you sure? I really need to get on that flight, my friend."

I slid the piece of paper off the bill so that it was clearly visible. The young man wasn't budging.

"Look, if you'll help me out, I'll give you a hundred dollars," I pleaded, in a desperate attempt to avoid being inconvenienced.

"Sir, I'm sorry. It's too late. I'll have to re-book you on the morning flight. And there's no need for money. It's my job."

Wow. This gentleman was a professional who took pride in his work. He wasn't interested in taking a bribe. *Was I still in Afghanistan?* Someone call CNN and have them do a report on this developing story. While I was pissed off that I couldn't get on the flight, especially since I knew the plane was still there, I was very impressed by the whole process. I had gone through a thorough search procedure with no problems, and I didn't even speak the language. Everyone was courteous and took their job quite seriously.

●●●

I pondered sleeping in the airport, but had a better idea pop into my head. I would see if any rooms were available at the Gandamack Lodge. That's where all the reporters stay. It's a nice guesthouse and is owned by a former British reporter. It has a nice garden, which serves as the dining area. The security is good, the rooms are clean, and they have free Internet. A quick call to the lodge confirmed a room was available. The manager would be on his way to pick me up. My problems were quickly solved. I would also have a shot at being able to hook up with my friends Rob and Mullah Paul, and catch up on things over a few beers. I really wanted to run the progress of the book past them as well.

I checked into my room and took a quick shower. I called Rob, but unfortunately he was out in Herat with Abe. However, Paul was in Kabul and should be around somewhere. Paul answered the phone and the typical greetings were exchanged.

"This is Bad Blake," I said.

"I've played drunk, broke, divorced, and on the run…Bad Blake ain't never missed a show," was the reply.

We both loved that movie (*Crazy Heart*). Paul had given it to me with the assurance that it was my type of show. After watching it, I have predicted that I will end up much like Bad Blake, the main character. It has already begun, actually. I'm divorced, usually drunk, broke, and after I publish this book, I probably will be on the run. Paul said he would take a shower and be right over. He obviously knew the place. Thirty minutes later, I was chatting with my good friend. I showed him the concept for the cover of the book to see what his response would be. He loved it.

"How could you not want to buy this book? Just look at the craziness of the front cover," Paul offered.

"Are you sure that you're good with being on the front?" I asked.

"My man, I'm good with it."

I read a few excerpts to Paul, who spent most of the time laughing. While that was a good sign, he was obviously a biased critic. I agreed to mail him hard copies of the individual chapters for editing. I would not send him the file via e-mail, especially since he was in Afghanistan. It would inevitably be picked up by Carnivore, Echelon, Able Danger, or some other new catch-all data mining program being used by our intelligence folks. I didn't want to give up the ghost before I was ready and had a finished product. They could critique the work via buying a book, instead of merely intercepting the file.

We sat in the garden at the only available table. We were surrounded by tables full of journalists, contractors, and personnel from the various embassies. The setting could have been from any upscale restaurant in the United States. No problem with getting a beer, a stiff drink, or a big fat piece of bacon on your burger. War was something that was going on far away from the minds of these folks.

We discussed what had been happening since I left. Paul advised that the program had continually become more restrictive. If anyone left one of the prison camps, they had to be in an up-armored vehicle. Up-armored vehicles weren't something they had, so in effect, the program was being choked. The only teams who were running operations were Jalalabad and Kabul, because they were simply ignoring the rules. Everyone else was locked down on the prison camps, soaking up $250,000 U.S. dollars plus, of taxpayer money.

Paul elaborated he had tentatively been en route to the Kandahar team to wreak some havoc, but politics intervened and stopped his transfer. Apparently, word had gotten around that he and I were *a couple of cowboys,* and were *reckless.* Two of the advisors in Kandahar had called Paul on the phone, and confronted him about coming down there.

"We don't need any cowboys coming down here to Kandahar. It's dangerous enough already," cried one of the two gentlemen involved, who we'll refer to as Mike.

The second advisor got on the phone and chimed in as well. We'll refer to him as Jerry.

"It's dangerous down here. We don't need you coming down here trying to tell us what to do."

That statement was particularly comical to me. Those two hadn't run a single case. In reality, they needed Paul to show them how to handle business, and how to quit being afraid of their own shadows.

I had to laugh out loud when Paul told me about this. Coincidentally, I had been in the lead position when Jerry arrived in country. He was scared shitless, nervous, and exhibited cowardly traits from the first minute I laid eyes on him. He's a short guy like me, but all bulked up from too many steroids and hours in the gym. He's one of those who's so self-conscious, that he shaves his arms. I'm sorry, but if you are a grown man and you shave your arms, I have to consider you to be gay. That would be the simplest explanation. If you are gay, then I understand why you shave your arms and attempt to look feminine. It makes sense and there is a logical reason behind it. Again, I don't have anything against homosexuals, nothing at all. This opinion is not meant to be derogatory toward my gay friends. It is merely a statement of logic and I think they'll agree with me. If you're a heterosexual man and you shave your arms, you've got some serious self-esteem issues going on inside.

The entire time I spent with Jerry, he never got off of the cell phone kissing his wife's ass. It made me sick listening to his crying, babying, and sweet-talking her nonstop. It was annoying at the least, but quickly became unprofessional and interfering.

On one occasion, I had brought him with me to a series of meetings, just to take him under my wing. He was crying to his wife on the phone the entire time he was in my company, even as we were walking into a secure building. I had to tell him to get off the phone so we could go inside. We couldn't bring our phones into the secure operations center, so I told him he had to leave it in the cubby-hole outside. He looked at me as if I had shot his puppy. I was thinking, *Relax, motherfucker. You came here to do a job. You can cuddle with your wife after hours, on your own damn time.*

I couldn't wait to ship that coward out and away from me. If I had possessed any command and control at the time, I would have bounced him back to the States on the next flight leaving Bagram. He was a liability to the other men and a liability to himself. He was one of those who had to be there, only because he had no other choice. Through bad financial decisions, I'm sure—he had to take the job. He certainly wasn't there to take any risks that would warrant him receiving combat pay.

As far as Mike was concerned, I was surprised that he was involved in the crying. I had always considered him the calm professional who held the Kandahar team together. I erroneously thought he was a soldier. Apparently, the two had even taken their complaints to the manager in charge of the southern portion of the country. I was even more shocked at his statements.

"Old Don basically said that he didn't need a cowboy coming down there, and that he didn't want to deal with me *going native* like Blackard," Paul explained.

Going native? First of all, in that context, it's a discriminatory and racially biased statement. I was removed from my position because I was living among the Afghan people. That is a fact I proudly concede to. However, the Afghans are people just like you and me. They do not deserve to be referred to as natives.

...*If I were the only white guy living in a black community* in the *United States, and someone referred to me as "going native," the NAACP would have a field day with the person who said that. What if I were living on an Indian reservation? Would I be considered "going native" there as well? These types of statements in the U.S. could cause a person to lose their job in law enforcement, and could be construed as racial discrimination. Why was it OK to talk like that in Afghanistan?*

What really killed me about that statement was from whom it originated. Don was the recruiter who hired me and gave me a chance at making a difference in Iraq and Afghanistan. I owe him for the adventures of the past three years of my life. I am a subject matter expert in counter-insurgency because of the opportunity afforded to me by Don. He was a former LAPD cop who had done time on the mean streets of America, Iraq, and Afghanistan. I considered him a true warrior. He was a cop's cop. Emphasis on the word *was*.

I considered him a friend, and had even vouched for him to many people who had inquired as to his reputation. Years before, Don and I had shared more than a few beers together in D.C. Throughout my years of being a cop, I had learned to never do battle on the streets with a person unless I had done battle with them in the bar. That's the only real way to know if they will have your back when things get bad. I had done battle in the bars with Don. As a matter of fact, on one occasion the two of us and another friend got thrown out of a bar for our craziness. We weren't going quietly, and a brawl almost broke out. I remember a young kid pulling a knife and I think the owner of the bar had a gun. My buddy Bill almost knocked out the owner before we got him out the door. As the beat cops were on the way, we were narrowly successful in escaping and evading arrest. Bill hauled ass on foot as Don and I escaped in his car. Don drove me back to the hotel. That's the only part I recall.

The next day, Bill elaborated on what had happened next. Bill had beat feet out of the place and was humping it toward the hotel through the woods. It obviously took him a lot longer to reach the entrance because we were in the car. Bill said that as he was about to open the door, he looked down and saw me passed out cold in the bushes. I was stretched out like a dead dog, feeling no pain. Bill, being the large frame that he is, threw me over his shoulder and deposited me in my room. He even showed up the next morning and made sure I got up.

Incidents like that build camaraderie, and let you know who you can depend on when the shit hits the fan. All of those guys who got hired with us and never left their hotel rooms to share a few beers—I wouldn't trust them for one minute on the streets of Fallujah or in the mountains of Afghanistan. I would go to hell and back with Bill to this day, for digging me out of those bushes.

● ● ●

Don is a very good person. I'm going to explain what happened to him so that I now have to revoke his membership to the player's club. This is a lesson in sociology as well, which supports the theory that you are a product of your environment. After his last Afghanistan tour, Don took a job in D.C. After a year of day-in and day-out indoctrination into the company line, he had lost the sense of responsibility to accomplish the mission. He had been transformed into a company guy who was merely concerned with billing the government, keeping everyone safe, and not making any waves. He had gone from the mentality of having *big cases and big problems,* to the mentality of having *no cases and no problems.* It killed me to think that a true soldier had surrendered to merely riding the government gravy train. He had been transformed from a *true patriot,* to a *corporate patriot.*

Paul basically elaborated that the whole program had changed. No real cases were being generated, due to the environment that had been fostered after I left. The company just wanted you to relax like the rest of the collared-shirt mafia, and soak up that big money.

● ● ●

The real problem with defense contractors, like government employees, is that they are not paid based upon performance or results. Most contracts merely require that the company provide the personnel. An applicant has to meet pre-requisites, which are often stretched, but after they are hired they aren't pressed to do anything. The companies just want a warm, qualified body (qualified only in case the company's files get audited) to deploy down range in order to begin the flow of money.

My personal opinion is that many people who were hired on this particular contract are guilty of theft and hypocrisy. As cops, they would have certainly taken someone to jail for shoplifting a hundred dollars-worth of merchandise. But somehow, it was OK for them to deploy to Afghanistan, lounge around on the prison camps, and collect over $250,000 U.S. dollars per year for their laziness and inaction. It's apparently OK to steal $250,000 U.S. dollars of the taxpayers' money, but NOT OK to shoplift a candy bar at Wal-Mart. You'd have to go straight to jail for that. The company is even more complicit, because they condone this behavior. They would love to have nothing but lazy people employed. Why? Because lazy people don't make any waves. *Just sit over there on the prison camp, be quiet, send in your fabricated situation report and skewed statistics, and we all get rich. Only the taxpayers get poorer, but that's not our concern, nor should it be yours.*

● ● ●

I have constantly argued a couple of points to people within the contracting business and the U.S. government. The first one is that we weren't getting paid a quarter of a million dollars a year because we were smart. If that was the case, we would have been making that much money while living comfortably in the States. The taxpayers were providing us with big money because there were risks involved with doing the job. You were well compensated for the danger and for the risk, up front. Therefore, you (in good conscience) could not utilize danger and risk as an excuse not to go to work or to leave the prison camp.

Secondly, defense contractors as a whole, Company X included, refer to the U.S. military as *the customer.* That is absolute arrogance if either Company X, or the military, think that the military is the customer. The true customer is the United States taxpayer. It's their money. Therefore, they are the customer. A customer expects to receive a return on his investment. Everyone has lost sight of that reality, which explains why neither is concerned with producing tangible results.

● ● ●

It was nice to be able to talk to Paul again. He's a stand-up guy and a true friend. As night crept into morning, we had to part ways. I had to catch an early flight back to Dubai. While it wouldn't be the last time we hung out, it would be the last time we met in Afghanistan. The band had been broken up.

As I transited through security at the airport the next morning, my bags ended up going through three different magnetometers before getting to the departure lounge. On this occasion, I had plenty of time to spare and found myself waiting with the other passengers.

We boarded the plane as scheduled. Again, I was impressed by the efficiency of the airport and its employees. The Germans had really done a great job in training the security staff. The whole process had been smooth sailing. As the ground crew backed the Boeing 767 away from its parking spot, I started to relax and get settled in.

A few seconds later…*BOOOOOOOM!!!!* The plane came to an abrupt stop, which threw me backward in my seat. My bones were briefly rattled. Everyone on board began to look around, to see where the indirect fire had hit. But sadly, it wasn't an incoming mortar round or a 107mm rocket. It was worse. The ground crew had backed the wing of the plane into the building. *Whoops.*

I guess the Germans hadn't focused on that part of the training. My beaming confidence in Afghan airport operations had to be put on hold. Rome wasn't built, or in this case, *rebuilt,* in a day. As we boarded a backup Boeing 737, I had to laugh. Things would only get better for the Afghans. Backing a plane into a building was merely growing pains. Accidents like that happen, even in America.

This trip psychologically ended my desire for further involvement in *the war* in Afghanistan. I would only return to Afghanistan as a tourist, or in a capacity not associated with the U.S. military. Maybe I will spend every Christmas in Afghanistan with my friends. I will always retain the memories of my time with Sammy and our team. The risks and the craziness of our actions will live on—only in this book, through pictures, and in the memories of a select few.

AFTERMATH

Drunken Advice

Recommendations Directed to the White House

Without delay, withdraw the majority of U.S. military forces from Afghanistan. Leave behind only non-conventional mindsets, of a more mature and refined variety. The young and aggressive with a *kill them all* mentality have no place among the Afghan culture. Those with that type of mentality are merely inflaming the situation and have no respect for the people. Older U.S. Army Special Forces soldiers should be the only ones on the ground and need to be allowed to operate under their own rule. Conventional-thinking military commanders hinder their efforts, and should immediately be removed from their chain of command and influence. The leaders in the Pentagon will never bow to this unless an executive order is issued to circumvent the bureaucracy.

A troop level of 5,000 is more than sufficient, as long as you have 5,000 people who are there to work and who possess the right attitude. When the President approved the increase in troop levels, it was a futile effort and only put more young American kids in harm's way. I don't blame the President for his decision, because he was only acting on the information presented to him. He does not hail from a military or police background. Therefore, he has to rely on his advisors.

In reality, we didn't need 30,000 more people to come to Afghanistan and sit on the prison camps. This equated to more convoys having to travel the roads to deliver the mail, ice cream, and gym equipment. The increase in convoy traffic created the opportunity for more IED strikes, which resulted in more American kids getting killed and injured. We already had 60,000 troops in Afghanistan who weren't doing anything truly productive. At this point, I believe 5,000 troops are sufficient to continue the mission of enabling and training the Afghans. It reduces our presence and will allow the Afghans to return to a sense of normalcy.

Immediately, cease the Central Intelligence Agency from flying drones into Pakistan. For that matter, put an end to any and all drone attacks conducted by the CIA. This basically amounts to a death penalty case, which is tried and adjudicated on less than sufficient evidence. It's not conducive with our values, laws, and principles as Americans. Half the time, they don't even know who they are killing. That's why you will read that "twenty militants" were killed, rather than a roll call of the names of the deceased.

For every person who is killed by the CIA during their misguided efforts, one hundred others are inflamed, enraged, and begin to hate the United States. The CIA is breeding new U.S. flag burners every day. Poor, uneducated people do not understand politics. They only understand that the Americans killed their family member. This will also take an executive order, because every time they fire a missile, someone's stock goes up. CIA personnel think they are above the law, so I don't believe what anyone has to say will actually alter their operations. Regardless, they are a big part of the problem in the region, and are far from being the solution. It is better to be thought a fool than to open your mouth and prove it. In this case, it is better to be thought of as evil, than to have the CIA kill a bunch of innocents, and erase all doubt in the minds of the population. The CIA is chock full of scholars, that's a given. However, they are what the old folks used to call *educated fools*. They have plenty of book sense, but no commonsense. They definitely have no idea of how to deal with poor, uneducated people.

Some will view this thought process as unpatriotic. *The lives already lost will have been in vain if we just withdraw,* is what will be cited. I understand that position, and argue that there is just no need to lose any more. Those who gave their lives made the ultimate sacrifice for our country. Many were awarded medals posthumously for their heroism. But, any one of their family members would piss on those medals if they could get their loved one back. Just ask them, and listen to their response. Ask a grieving mother if she truly believes her son's death was necessary.

● ● ●

Much like the *War on Drugs,* the conflict in Afghanistan is a buffet program. It's all you can eat for as long as you want to eat. We could send 200,000 more troops for another ten years. Insurgents would still be attacking the FOBs and convoys in the year 2022. The time to end this is now, over all objections by the private companies who are making a killing at the expense of the U.S. taxpayers.

Ho Chi Minh reportedly said, "If the Americans want war, we will give them twenty years of war. If they want peace, we shall invite them to drink tea this afternoon." We didn't consider this offer. We insisted on going to war. We lost. But as I write this, U.S. Navy vessels are currently in the Port of Da Nang, conducting training with the Vietnamese. We could have arrived at the same outcome, without having to kill two million Vietnamese civilians.

Regardless of your opinion of the Vietnam War, the situation in Afghanistan is the same. Had we stayed in Vietnam for another ten years, we would have lost another 58,282 young American kids. If we stay in Afghanistan for another two years, we will only have more grieving families to show for the effort.

●●●

Whether we want to admit it or not, in the near future, the Taliban will be a recognized political party in Afghanistan. They may even return to power. That's reality. It might behoove us to start showing them a bit of respect, now. Maybe not love, but at least respect if we want to coexist with them in the region. Remember, there were no members of the Taliban on the planes of 9/11.

And by the way, have we forgotten that some of the Taliban (former Mujahideen) were the ones directly responsible for the beginning of the collapse of the Soviet Union, and thus the end of the *Cold War?* They spent ten years doing our dirty work by challenging the most powerful enterprise on earth, mostly on foot or on horseback. The *Cold War* had the potential to annihilate the entire human race, and not just a few thousand people. Now, we are killing them by the droves, because the U.S. government's public relations machine has convinced the American public they are our enemy. In my opinion, they are the least of the threats to our country. While we are spending a billion dollars a day, trying to chase a bunch of farmers around with million-dollar high-tech gadgetry, there are more pressing and dangerous issues being neglected. For example, our own infrastructure.

There are bridges in the United States that are unsafe, roads that need repairing, electrical grids that need upgrading, and dams that need attention. Our crumbling infrastructure has the potential to kill a hell of a lot more people here in America, than foreign terrorists have ever thought about. Those facts remain hidden from the American people.

Our nuclear facilities can never be upgraded enough to ensure our safety. Reportedly, in March of 2006, thirty-five liters of a uranium solution leaked at the nuclear fuel services plant in Erwin, Tennessee. The facility was shut down for seven months. In November of 2005, nuclear material was found to be leaking into the groundwater near Exelon's Braidwood Station, in Braidwood, Illinois. The last reactor incident was in 1979 at Three Mile Island. Statistically speaking, we are due for another major mishap. The billions wasted in Afghanistan could be re-directed to reactor improvements, especially in light of what just happened to the Japanese.

● ● ●

Of course, all of these recommendations are irrelevant if one buys into any of the conspiracy theories I learned from my Afghan friends. The one that immediately comes to mind, is the real motivation behind the Russian invasion of Afghanistan. This was brought to my attention one day when we were on the way back from Torkham Gate.

There is a waterway system which was built out in the middle of nowhere. I was informed that the Russians built it so that several remote villages could have water. It was an elaborate system that obviously cost someone a lot of money. Now, why in the hell would the Russians waste their time supplying water to remote villages, when they were killing the people anyway? The answer provided was that the waterway was constructed in a manner to conceal the mining activities of the Russians in the mountains. The mining of uranium, which was the real reason they spent ten years in Afghanistan taking a constant beating. They were killing the villagers to keep them at bay, and to protect the integrity of the projects.

Therefore, could it be that the reason we have spent so much money in Afghanistan, is because of future plans to mine uranium? I will concede it's definitely conspiratorial without any supporting evidence, but it is an entertaining theory, nonetheless. It would explain why we invaded a country that had little to do with the attacks on 9/11.

Maybe it had something to do with the proposed pipeline project from Tajikistan to Pakistan, that a Texas-based company was trying to win, while competing with a company in Argentina? I could go on and on, but you get the point. Those theories might explain the situation and the actual motivations for being there. Maybe, it's just the fact that there are trillions of dollars of other precious minerals waiting to be mined in those mountains.

● ● ●

My final recommendation would be for President Obama to listen to the sermon presented by Doctor Martin Luther King, Jr., on April 30th, 1967, at the Ebenezer Baptist Church in Atlanta. The sermon addressed the issue of why Dr. King was opposed to the war in Vietnam.

Dr. King began by saying, "The time has come for America to hear the truth about this tragic war. Now I've chosen to preach about the war in Vietnam today because I agree with Dante, that the hottest places in hell are reserved for those, who in a period of moral crisis, maintain their neutrality. There comes a time when silence is betrayal. The truth of these words is beyond doubt, but the mission to which they call us is a most difficult one. Even when pressed by the demands of inner truth, men do not easily assume the task of opposing their government's policy, especially in time of war."

Dr. King's references and views on the Vietnam War could easily be transposed to the war in Afghanistan. Dr. King referenced an estimate suggesting that it cost U.S. taxpayers five hundred thousand dollars to kill one enemy soldier in Vietnam. Meanwhile, only fifty-three dollars were spent for each person classified as being poor. I would surmise that today, we spend fifty million dollars per every Afghan farmer we kill. I would argue that the money could be better spent, on a more noble cause—a cause that would actually benefit the American people and not just the one percent of the rich who control the beltway.

Recommendations Directed to the American People

We must wake up and take note of what has happened to our country since World War II. President Eisenhower warned us about what he termed the *Military-Industrial Complex*. Reportedly, he first wanted to call it the *Military-Industrial-Congressional Complex,* but political pressure forced him to drop the word *congressional.* Regardless, in my personal experience and research, I believe the more accurate description is the latter. I refer to it as *MICC,* for short. Most Americans think this term is something that was conjured up by conspiracy theorists, communists, nut cases, or even Michael Moore. It was not. It was coined by our former

President. What is particularly special about President Eisenhower, is that he was also a general in the U.S. Army before being elected. That made him a subject matter expert on the U.S. military, as well. Was he crazy? No, he wasn't crazy. He was no conspiracy theorist. What was he? He was informed. He was well informed. He was concerned with what was happening to our country because of this convenient relationship. Imagine the courage it took for a former General to speak out against his own organization. His affiliation with the U.S. Army was what had catapulted him into the White House in the first place.

Even before President Eisenhower spoke of the *Military-Industrial-Complex* during his farewell address, he had made several other valid points during his presidency. On April 16th, 1953, shortly after taking office, he gave a speech entitled *The Chance for Peace,* which was delivered before the American Society of Newspaper Editors. A few excerpts from his speech are worth noting. He contrasted the views of the Soviet Union to that of the United States, by saying, "The way chosen by the United States was plainly marked by a few clear precepts, which govern its conduct in world affairs. First: No people on earth can be held, as a people, to be an enemy, for all humanity shares the common hunger for peace and fellowship and justice. Second: No nation's security and well-being can be lastingly achieved in isolation but only in effective cooperation with fellow nations. Third: Every nation's right to a form of government and an economic system of its own choosing is inalienable. Fourth: Any nation's attempt to dictate to other nations their form of government is indefensible. And fifth: A nation's hope of lasting peace cannot be firmly based upon any race in armaments but rather upon just relations and honest understanding with all other nations."

The third point basically says that countries can determine how they govern themselves. In Afghanistan's case, the Taliban was the constituted government. In Iraq's case, Saddam Hussein was the government. The fourth point basically says that there is no excuse for an outside nation to impose their views on that of another. If that is supposedly one of our precepts, then why does the United States attempt to impose democracy on every nation in the world? Sure, as Americans, we think democracy is the best thing for everyone, especially since we're always right.

President Eisenhower went on to say, "...Every gun that is made, every warship launched, every rocket fired, signifies in the final sense, a theft from those who hunger and are not fed, those who are cold and are not clothed. This world in arms is not spending money alone. It is spending the sweat of its laborers, the genius of its scientists, the hopes of its children. The cost of one modern heavy bomber is this: a modern brick school in more than thirty cities. It is two electric power plants, each serving a town of sixty thousand population. It is two fine, fully equipped hospitals. It is some fifty miles of concrete pavement. We pay for a single fighter plane with a half million bushels of wheat. We pay for a single destroyer with new homes that could have housed more than eight thousand people. This is, I repeat, the best way of life to be found on the road the world has been taking. This is not a way of life at all, in any true sense. Under the cloud of threatening war, it is humanity hanging from a cross of iron." The reader should remember that the figures quoted were in 1950's dollars. In today's currency, the numbers would be staggering.

We, as Americans, have ignored President Eisenhower's warnings. Because of this, we have allowed the leadership of this country to act like imperialists, and drag us into conflict after conflict—conflicts that have continued to bankrupt our nation. Conflicts that only serve the purpose of fueling the *Military-Industrial-Congressional Complex,* and maintaining certain stock prices.

Dr. King referred to the complex as *militarism*. I refer to it as *imperialism*. I'll allow you to look up the definition of imperialism for yourself, but it basically means *fucking* with other people. As an American, I am tired of my tax dollars being siphoned for the purpose of useless violence. I don't want to spend a billion dollars a day to kill a farmer, seven thousand miles away in Afghanistan. I would rather spend a billion dollars a day to build schools or repair roads, right here in America. We need for those who inhabit the beltway to stop wasting our money on things that do not benefit us.

The single biggest threat to our nation right now is our national debt. This debt continues to grow, largely due to *MICC*. Under the guise of what Dr. King called *smooth patriotism*, the rich have convinced the American public to wave their flags, don't question the government, and support all military action. That's how the rich get away with it. The problem is, they are digging a hole today, that our children and grandchildren won't be able to get out of.

In regards to Afghanistan specifically, the time is now for congress to introduce a Case-Church Amendment, to force an end to the madness. It wasn't popular with the defense contractors in 1973, nor will a similar amendment be popular with the defense contractors today. I know this suggestion will be looked upon as absurd, and in reality, it is. Why?

Because since the passing of the Case-Church Amendment, congress has become so intertwined with the defense industry, there is no longer a definitive line that divides them. We could begin to refer to congress as *MICC, Inc.,* since they have essentially become a business, out for profit.

Recommendations Directed to the Pentagon

If the President continues to order you to deploy troops to Afghanistan, or other nations involved in an insurgency, then evaluate a few thoughts.

Realize your limitations. Believe it or not, you actually cannot accomplish every mission with your current tactics, techniques, and procedures. Counter-insurgency is one mission that comes to mind. Go back and study Vietnam history, and come to realize we haven't learned a thing. We lost in Vietnam. We did not succeed in Iraq. (We had to start paying people not to fight us, and then get the hell out of there. That's not exactly the definition of success.) We aren't winning in Afghanistan. Ask yourself, why? America had, and has, more equipment, weapons, and gadgetry than the Vietnamese, the Iraqis, or the Afghans ever thought about. Why couldn't we quell these insurgencies? I have two opinions to offer as an explanation.

First of all, in the U.S. military's defense, I don't think an invading force can ever win at counter-insurgency. As an invading force, you are trying to liberate people who don't want to be liberated. Mostly, they just want you to go home and leave them alone. They view you as imperialists, sticking your nose where it doesn't belong and interfering with their internal affairs. It causes the local citizens to temporarily put aside their differences, long enough to band together and fight the foreign invaders who are viewed as the bigger threat. They want to handle their own problems, which is understandable.

It's not as if we are invading France in 1944, where the population wants to be liberated. That mission was clear and the population was with us. But in Vietnam and Iraq, we were trying to convince the population that U.S. government policy was what was best for them. You can't even convince me that U.S. government policy is what's best, and I'm an American. The situation in Afghanistan is no different. We are trying to liberate people who don't want to be liberated. They just want us to leave. What makes you think you can convince a poor, uneducated Afghan that he needs to do everything the Americans tell him? And, why should he?

Imagine if Country X had tried to intervene during our Civil War and dictate what was best for America. What do you think would have happened? The North and the South would have temporarily banded together and fought Country X. Once Country X was expelled, the Confederacy and the Union would have gone right back to fighting—similar to what's happened in Iraq and what will happen in Afghanistan.

In a nutshell, an outside military force cannot win at the game of counter-insurgency. Local people will resent foreign military personnel, no matter what uniform they wear. For some reason, the United States cannot seem to understand, nor accept this concept. Often, the U.S. appears to be the *only* nation that cannot comprehend this.

Conventional military forces are not the right tools for the job. The Central Intelligence Agency and the State Department, from the U.S. perspective, should be the right tools for the job; but, they can't seem to figure it out, either. However, counter-insurgency operations should fall to them, and not the military. Unfortunately, they are staffed with too many scholars and *educated fools,* and not enough practical thinkers.

Insurgency can, and may, be expressed through mere violence by those who initiate it. That's very easy. *Counter-*insurgency is a game of wit, negotiations, relationships, deals, and alliances. Violence, as a strategy of counter-insurgency, only energizes the insurgency even more. It unites the insurgents and solidifies their cause. What's that saying? *Remember the Alamo?* Go dig up old Santa Anna and ask him how his counter-insurgency strategy worked out.

Second, the reason the U.S. military does such a poor job during counter-insurgency operations is because the wrong mentality is applied. The mission is doomed before the first boots hit the ground. Personnel are trained to *stay in their lanes* and are not permitted to *think outside the box.* It's like trying to create abstract art on a canvas full of grid lines. It doesn't work. Conventional thinking and conventional warfare equates to playing a game of checkers. It's a simple game that most everyone can play. Conducting counter-insurgency operations requires unconventional thinking and methods, which equates to playing a game of chess. Not everyone can play the game, especially conformists.

Think about what the word *counter-insurgency* really means. What it means is instilling confidence in the people of the local government's ability to govern. It means enforcing the law of the local government. It does not have anything to do with killing people, dropping bombs, calling for fire, Predator strikes, or the named operation you thought up while taking a shit. Those activities have to do with conventional warfare, your personal quest for medals, and your fitness report. Counter-insurgency has nothing to do with your/our opinion as Americans, our views, or what the hell we think. It has everything to do with what the local government thinks, and how the local government chooses to govern its people.

What works in the States does not work in Afghanistan, from policy to procedure. What works on paper and in briefings does not work on the ground. If mission accomplishment is always 51% over every other issue, then it should trump all internal formalities, rules, and other useless regulations that serve a purpose while in garrison. You have to allow soldiers to flex to the environment.

Money and force will never change a person's mentality or beliefs. As Americans, we think we can buy anyone's friendship. We cannot. Threats of violence do not change culture, either.

An old sheik in Fallujah, Iraq, made the statement, "You Americans need to take a lesson from the British. They know how to talk to people and how to negotiate. You Americans come in and demand things, *immediately*. If you don't get your way, you either throw down a bunch of cash or resort to violence. The British can merely talk to people and get what they want."

Take the mounds of cash away from the Central Intelligence Agency and they would be useless, because they have no people skills. This explains why they say you can rent an Afghan's loyalty, but you can't buy it. Well, no shit. You give me a box of money and sooner or later I'll be wanting another one, because I have no personal connection to you. It's merely business.

● ● ●

Ban the use of PowerPoint for any unit or soldier below the Brigade level. Soldiers should be interacting with the Afghans or the locals wherever they are, instead of trying to impress one another with statistics and graphic design skills during five useless meetings a day. Name one military leader who is remembered for his or her PowerPoint skills or e-mail etiquette. There are none. No great leader will ever be recognized, awarded, or remembered for his PowerPoint skills or his e-mail etiquette. These qualities have nothing to do with leadership and cannot be used as a measure of performance in the business of counter-insurgency. Bill Gates is rich enough already, so he probably won't mind the restrictions.

Put your civilian experts to work. If they don't want to work or are incompetent, then fire them. The same applies to your interpreters. Hold people accountable for a day's work. They're making more money than you anyway, so empower them and put their ass to work instead of suppressing them. If you suppress them, they'll be glad to get paid for doing nothing. They'll have no problem soaking up that combat pay while lifting weights in the gym. If you enable them, they will make you look good and you can take credit for their work.

Get rid of the steroid powder, steroid shakes, steroid candy bars, supplements, pills, etc., and the gyms. Getting all *roided up* to impress your girlfriend or wife back home has nothing to do with the mission. It is part of the reason we've been in Afghanistan so long and accomplished so little. Any man caught shaving his arms should be considered underworked and given something to do. If he has time to shave his arms, he isn't concerned with the mission. Pull-ups, push-ups, sit-ups, and interacting with the Afghans is plenty enough physical exercise to keep soldiers in shape. If they complain about not having a gym, they're not doing enough. Afghanistan has become solely a year-long workout plan for ninety percent of the troops deployed there.

Stop wasting soldiers' time counting bullets and toilet paper. More time is wasted conducting inventories of items that should be considered expendable anyway. It detracts from the soldiers' ability to actually accomplish something. In a deployment, things are going to get lost or broken. Millions of dollars are wasted on a daily basis in Afghanistan. So, why discipline a young nineteen-year-old PFC when he loses his night vision goggles while on a patrol? That's a great idea. Let's kill this kid's career, along with his commanding officer's, over a thousand-dollar piece of equipment. Meanwhile, we just spent three hundred thousand dollars on new equipment for the gym. Commonsense needs to prevail over counting beans and creating spreadsheets.

In what you refer to as a COIN environment (trying to make friends and instill confidence in the people of the current local government), lighten up. Take a platoon of soldiers and have them move into any neighborhood in America. Initially, they will be welcomed. But after a week, even the American people will grow tired of their activities and wish for them to go. The Afghans are no different. Imagine long, lumbering military convoys driving through your neighborhood every day, blocking traffic. How do you think the Afghans feel?

Realize the fact that the Afghans don't work for you, even though you perceive yourself to be the *battlespace owner*. That's arrogant and inaccurate. It's their country and Afghanistan is not America. Show respect for that. You are a guest. If you act like a guest, they will treat you like a guest. Act like an invader, and you will get treated like an invader. *Did I mention the Afghan community does not answer to you? Try not to forget that concept.*

Delegate some real authority to that young company commander out in the hills. In turn, he needs to let his young lieutenants make actual decisions. Above their level, ground truth ceases to exist. Anyone above the rank of captain has no clue as to what's really going on in Afghanistan, even if they are in Afghanistan. If you're inside the beltway, you certainly don't have a clue as to what's really going on there. You can trust me on that one as well. I don't care how many intelligence reports from the CIA you've read.

Get rid of the body armor. Soldiers can't chase bad guys, support the Afghan security forces, or interact with the public while wearing a hundred pounds worth of gear. If the prison camp gets attacked, then by all means feel free to put it on while manning the wall. You'll feel safer when a bullet hits you in the head. But, when you're on patrol, you're basically too fat to fight. Sure, the contractors who manufacture body armor will lobby to fight this policy change because they might lose money or their stocks might dip. If no one is wearing body armor, it might give the perception that no one is afraid. No fear, no money. I went an entire year on two sets of man-jams, an AK-47, and a blanket. That's all the bad guys have as well. The toughest job your soldiers face is getting from point A to point B with the ton of junk they have to drag with them. Lighten the load.

If a fight does happen to come your way, stop running from it. You've just accomplished what the entire intelligence community cannot seem to do—find real bad guys. Capitalize on the find. Have young soldiers dismount, give chase, and hunt them down. It's quite alright if they get out of gun range from the MRAPs. They can hold their own. The bad guys are armed with rifles, machine guns, RPGs, and grenades. A U.S. dismounted patrol is armed with about the same equipment. Only if they are severely outnumbered and no air cover is available, should they push (run) from a fight. It sets the wrong precedent and just encourages more attacks.

One thing that doesn't make any sense to me, is that we established some of the FOBs in old positions the Russians used. That defies logic. The guys who we now refer to as insurgents (former Mujahideen) had ten years of practice to perfect their attack on the same piece of ground. The minute we moved in, they started on decade number two.

The second thing that defies logic is the fact that a lot of these FOBs are located on low ground. Ridgeline surrounds some of the FOBs in the worst areas. Who had the bright idea of putting a fixed position down in a bowl? I do have a theory on the reasoning behind this situation. It's because MRAPs can't make it up a hill. Therefore, in order to get the mail, the ice cream, and all the other unnecessary bullshit (gym equipment) to the remote prison camp, it had to be located on low ground. Never mind the fact it resembles a shooting gallery from the surrounding terrain. I've sat on a ridgeline while looking down at a FOB and wondered, *Why in the hell did they put that thing there?* You need to close these death traps immediately and move to high ground. An ROTC student has enough commonsense to know this.

● ● ●

To qualify and contrast my dissenting opinion and perceptions, I would suggest you research the experiences of a gentleman by the name of Captain D.L. "Pappy" Hicks. Captain Hicks ran teams of Hmong tribesman fighting a secret war in Laos, during Vietnam. From what I've read about this American hero, he suffered from the same opposition, bureaucracy, and suppression that I did. He was a one-man show, running with his team under the radar. He had to sneak around to run ambushes with his men because the bosses thought, *You can't trust a gook.* I believe Captain Hicks chose to refer to the Vietnamese as friends and not *gooks*—the same way I refer to the Afghans as friends, instead of *Hajjis*. I don't want to speak for him, but I suspect his tales and frustrations would mirror mine. I believe he was a good man and many lessons should have been learned from his experiences. From my perspective, Captain Hicks had the closest recipe for success. Unfortunately, the brass-in-the-box (conventional thinkers) in Saigon, with their charts and graphs, continued to think that B-52s and body counts were the answer.

● ● ●

Finally, and most importantly, locate and court martial the *asshole* who came up with General Order Number 1. Whoever it was, is a real *fucking* idiot and should be punished for depriving our men and women for so many years. He certainly doesn't know *shit* about building relationships, mission accomplishment, nor keeping your people motivated. Then, abolish General Order Number 1 and build bars on every prison camp. That way, the young soldiers and Marines can invite their Afghan *friends* over for a drink, build some real relationships, and actually solve problems. Consider this to qualify that line of thinking. President Obama diffused a potential racial time bomb after a white cop arrested a black professor. How? He invited the two of them to the White House. For what purpose might I ask? It was for the purpose of discussing and resolving the situation over a beer. While many ridiculed the President for doing this, I have to applaud him. He took a heated situation, which had the potential to spark violence in the streets of America, and solved it with face to face conversation over a beer. It was solved with simplicity. You haven't heard anything about the incident since, have you? The incident speaks for itself.

Immediately, serve drinks on all freedom flights and don't preach to the soldiers about how much to drink, while laying over in Ireland. They already have a set of parents who have covered that topic. If they've just survived a year in Afghanistan, they have earned the right to get plastered for the ride home.

Aftermath and Fallout

Although I authored this book in 2010, I purposely delayed the publication until 2012. I still had some work to do with Company X which would benefit our young servicemen and women. I knew that once the book went public, it would mean my demise. I chose D-Day (June 6th) as the target date in which to let the cat out of the bag. I also decided to schedule some vacation time to coincide with the revelation. After dropping a rough draft in the mail to my boss Smitty, at Company X, I went off the grid for a couple of weeks. My buddy compared the pending opening of the package to the pulling of a pin on a grenade.

Smitty made several desperate attempts to reach me, but I had already left the country and was out of communication. The separation would give the company ample time to digest the contents of the manuscript, and take a position. Afterward, we could all convene and discuss the issues, amicably.

Two weeks later, I returned to my duty station in the States. I reported in at 7:00 a.m. and got right back to work. At 10:00 a.m., I was on the phone with Smitty. He advised that because of the manuscript, the company had placed me on leave, without pay. I was to immediately "retreat" from the work site. But, *Oh by the way, we'd like for you to come up to D.C. and discuss the book.* They would get me on a flight the next morning and provide hotel accommodations. Since I didn't have anything else to do, I told Smitty not to worry about a plane ticket. I drove straight to D.C. and checked into the hotel. The next morning, I reported to headquarters as requested. I imagined there would be an entire committee present, from attorneys, to marketing people, to bosses, etc. I thought they would express their concerns over certain elements of the book and request that I make some changes. I went into the meeting with the mindset that it would be a negotiation of sorts. I had packed a copy of the book and a red pen because I was prepared to make concessions for the good of all. My vision of the meeting was far from what was about to occur.

I arrived early and reported to the appropriate office. The receptionist ushered me to the conference room. Already present was a rather attractive lady sitting across the table. She was dressed very professionally, which tended to indicate she was an attorney, and indeed she was. She introduced herself and offered me something to drink. As we made small-talk, it was apparent she was regretting being early. She found herself having to entertain the opposition while waiting for the rest of her crew to arrive. Regardless, she was very nice to me and I found her to be a sincere person.

A short time later, Smitty's assistant (who we'll call Larry) walked in. His demeanor was not the least bit friendly. The only thing he said to me was, "Where did you go on vacation?"

I responded by saying, "I went a lot of places."

I laughed to myself at his poor attempt to elicit information. There was no small talk or cordial conversation—only the question. While it was none of his business to begin with, I wouldn't have told him just because he was rude. Had the nice lady asked me, I would have told her anything she wanted to know. I'm not sure, but I think Larry was trying to find out if I had gone to Afghanistan during my vacation.

The last to arrive was another attorney. He was less than cordial and a bit too cocky for his own good. As the old folks used to say, *He was too big for his britches.* He advised me that the purpose of the meeting was to conduct an investigation. *Conduct an investigation? This should be interesting,* I thought. He chaired the event and needless to say, we didn't see eye to eye. I won't reveal the issues, but would like to elaborate on a couple of points I found relevant.

According to the gentleman, he was a former Marine and a pilot. He's what infantry Marines refer to as an *Air-Winger*, which, from their perspective, means he's part of the softer side of the Marine Corps. In discussing certain events that occurred in the book, he tried to convince me that he too, had vast experience working with *local nationals*. He said the words *local nationals* in such a professional and erect manner that he reminded me of that asshole major from Mehtar Lam. The way he said it convinced me that he was lying, as well. He was purporting himself to be on my level of expertise when it came to dealing with the locals. However, he was a pilot. A pilot's perspective is from the flight deck and not from the level of ground truth. I've flown in many aircraft before, but I don't claim to be an aviator.

Early into the conversation, I realized that the packet of papers he was holding was a photocopy of an entire chapter from my book. Apparently, he had missed the copyright symbol at the beginning of the text. I'll admit that the *C* with the circle around it is a bit small. Maybe the guy doesn't have good eyesight from looking at all those books in law school. However, it is spelled out in black and white that no reproduction of this text is authorized, except in short literary reviews. Isn't he an attorney? Isn't the purpose of copyright law to prevent such acts? I suddenly felt as if I were the victim of a theft. I decided to call him out.

"Did you copy my book? Didn't you see the copyright symbol?"

"It's called fair use. Look it up," he responded, condescendingly.

I couldn't focus on anything else the guy had to say. He had violated federal law, from my perspective. As far as I was concerned, I was being questioned by a suspected felon. For some reason, I just kept thinking about Sean Parker and Napster.

In good faith, I had sent Company X a free copy of the book. They took the free copy and reproduced it. While I can only speculate as to how many illegal copies were made, I do know one thing. Someone at Company X owes me $21.95 for that bootleg copy the attorney was holding. I also have to assume that the manuscript survives on corporate computers as a PDF document as well.

After the gentleman and I beat a dead horse for an hour, he apparently got what he wanted. While he was doing his job, the only thing he was concerned with was protecting corporate money.

I wanted to ask this gentleman, *At what point did you add the word corporate to your title of patriot?* At one point in his life, he was a Marine—a true patriot. But, he had sold his soul to the beltway and the Military-Industrial-Congressional Complex. I can speak intelligently, because I was also temporarily part of the problem. As a matter of fact, everyone in the room was a former government employee who had left to join the *Beltway Bandits* and chase the big bucks.

The nice lady asked me about why I had sent a copy of the book to my boss. I responded by telling her it was the gentlemanly thing to do. It wouldn't have been right to just publish the book without warning. It was more appropriate to address the issue up front, rather than waiting until the bosses saw the book at Barnes and Noble. I really think they all three missed the point of why I had come to the meeting. I was there to get them to weigh-in and voice their opinion. I was prepared to make changes. I was prepared to negotiate the content.

The meeting abruptly ended and I was declared *persona non grata*. According to Larry, I wasn't being fired. I would remain on leave, without pay. I would ultimately receive no further communication until about a month later. An e-mail came from human resources which contained two attachments. The first attachment was a letter from the nice lady in the meeting. It basically advised that Company X would sue me if I published the book. *So much for her being nice.*

The thing that Company X doesn't understand, is that I'm not afraid of litigation. President Eisenhower spoke out against the military and congress, even though it wasn't popular and went against the grain. Dr. King went to jail and gave his life for his beliefs. What kind of person would I be, if I allowed the threat of litigation to scare me out of publishing ground truth history? I believe strongly in the First Amendment. I believe there is a reason that freedom of speech is guaranteed in the First Amendment, and not the second, third, fourth, etc. It's because it is the most important of all our rights. Any attempt to suppress this book should be viewed as an attack on the personal liberties of all.

The second attachment was a letter from human resources—my walking papers. As I read the first line, I could only laugh.

"*Dear Mr. Blackard: The company has determined that you violated the terms of GENERAL ORDER Number 1 as it applies to your employment in Afghanistan. Consequently, the decision has been made to terminate your employment from the company effective, Monday, July 23, 2012.*"

In the end, I got canned for drinking beer in Afghanistan. An alleged violation, which occurred over two years before. I'm not real sure what gave them that suspicion, though. It's such a perfect ending to the perfect adventure. What an appropriate thanks from Company X for the risks I took—risks taken in order for them to get rich. For that, I must say, *You are quite welcome.*

Acronyms and Initials

AAF	Anti-Afghan Forces
ACLU	American Civil Liberties Union
ANA	Afghan National Army
ANP	Afghan National Police
AUSA	Assistant United States Attorney
BATS	Biometric Automated Toolset
CAC	Common Access Card
CEXC	Combined Explosives Exploitation Cell
CI	Confidential Informant
CIA	Central Intelligence Agency
CID	Criminal Investigations Division
CIF	Central Issue Facility
CJTF	Combined Joint Task Force
COA	Course of Action
COC	Combat Operations Center
CONOPS	Concept of Operations
CONUS	Continental United States
CRC	CONUS Replacement Center
CSTC-A	Combined Security Transition Command – Afghanistan
DEA	Drug Enforcement Administration
ECP	Entry Control Point
EOD	Explosive Ordnance Disposal
FBI	Federal Bureau of Investigation
FOB	Forward Operating Base
GPS	Global Positioning System
HA	Humanitarian Assistance
HCT	Human Intelligence Collection Team
HUMINT	Human Intelligence

Acronyms and Initials

IDF	Indirect Fire
IED	Improvised Explosive Device
IO	Information Operations
ISI	Inter-Services Intelligence
ISU	Investigative and Surveillance Unit
JIEDDO	Joint Improvised Explosive Device Defeat Organization
LAPD	Los Angeles Police Department
LEC	Law Enforcement Contact
LIMDIS	Limited Distribution
LN	Local National
LZ	Landing Zone
MGRS	Military Grid Reference System
MOI	Ministry of the Interior
MRAP	Mine Resistant Ambush Protected
MRE	Meal, Ready-to-Eat
MSR	Major Supply Route
NCO	Non-Commissioned Officer
NDS	National Directorate of Security
NGO	Non-Governmental Organization
OGA	Other Government Agency
OP	Observation Post
ORSA	Operations Research and Systems Analysis
OSL	Off-Site Location
PHQ	Provincial Police Headquarters
POO	Point of Origin
PTSD	Posttraumatic Stress Disorder
PX	Post Exchange
RPG	Rocket-Propelled Grenade

Acronyms and Initials

SAF .... Small Arms Fire
SF .... Special Forces
SIPR .... Secure Internet Protocol Router
TCN .... Third Country National
TOC .... Tactical Operations Center
UXO .... Unexploded Ordnance
VBIED .... Vehicle Borne Improvised Explosive Device
WRA .... Weapons Reduction and Abatement
WSO .... Weapon Systems Officer

Almost forgot the most important acronym of them all...

MICC .... Military-Industrial-Congressional Complex

www.ingramcontent.com/pod-product-compliance
Lightning Source LLC
Chambersburg PA
CBHW032356040426
42451CB00006B/29